Applications of Circularly Polarized Radiation Using Synchrotron and Ordinary Sources

Applications of Circularly Polarized Radiation Using Synchrotron and Ordinary Sources

Edited by
Fritz Allen
and
Carlos Bustamante
The University of New Mexico
Albuquerque, New Mexico

PLENUM PRESS • NEW YORK AND LONDON

Library of Congress Cataloging in Publication Data

Workshop on Applications of Circularly Polarized Synchrotron Radiation (1984: University of New Mexico)
   Applications of circularly polarized radiation using synchrotron and ordinary sources.

   "Proceedings of a Workshop on Applications of Circularly Polarized Synchrotron Radiation, held May 18–20, 1984, at the University of New Mexico, Albuquerque, New Mexico"—T.p. verso.
   Bibliography: p.
   Includes index.
   1. Synchrotron radiation—Congresses. 2. Polarization (Nuclear physics)—Congresses. 3. Circular dichroism—Congresses. 4. Photoelectron spectroscopy—Congresses. I. Allen, Fritz. II. Bustamante, Carlos. III. Title.
QC793.5.E627W67   1984   530.4'1   85-19158
ISBN 0-306-42087-2

Proceedings of a Workshop on Applications of Circularly Polarized Synchrotron
Radiation, held May 18–20, 1984, at the University of New Mexico,
Albuquerque, New Mexico

. . . utilicemos la última luz para llenar los ojos
con tanta realidad abrumadora: . . .

Ángel González
Grado Elemental, 1962

FOREWORD

This volume has resulted from a meeting held May 18-20, 1984, in Albuquerque, New Mexico, to bring together people interested in all aspects of polarized radiation. The main emphasis was circular polarization. A very broad range of scientific disciplines was represented. It included the physics of synchrotron light sources, theory of polarized light scattering and absorption, experimental measurements using circularly polarized light from 40 eV (31 eV), identification of viruses using circular polarization, production and analysis of differential images using linearly and circularly polarized light. The interaction between theoreticians and experimentalists, between device designers and users, and between physicists, chemists and biologists was very stimulating. Everybody learned and everybody taught. Theoreticians learned that some of their assumptions were completely inappropriate for the experiments they were trying to explain; experimentalists found that some of their anomalies were easily explained by simple theory. It was obvious to all participants that use of arbitrarily variable polarized light will have an important impact on all aspects of structure determination on species ranging in size from atoms, to surfaces, to biological cells.

Even though polarization is a spectroscopic variable of radiation, as is intensity and wavelength, most spectroscopic experiments have ignored polarization, although as early as 1852 G. C. Stokes showed that four parameters were necessary to characterize light. They specify the intensity and the state of linear and circular polarization of light at any wavelength. To specify completely the interaction of light with matter (if only linear effects are considered) requires sixteen parameters that connect the four input Stokes parameters to the four Stokes parameters of the outgoing light. These sixteen parameters form the Perin matrix or Mueller matrix. Measurement or calculation of these parameters is necessary for a complete description of the (linear) interaction of light with matter.

The experimental research presented at the conference and reported here deals mainly with the visible wavelength region and slight extensions to either side (roughly from 150 nm to 1000 nm, 8.3 eV to 1.2 eV). A single exception was that dealing with a description of spin-resolved photoelectron spectroscopy at energies up to 40 eV (31 nm). This work was done using circularly polarized radiation emitted above and below the plane of the circulating electrons in a synchrotron ring. The device at BESSY (West Germany) in which the experiments were carried out seems to be the only one presently capable of providing circularly polarized radiation in the X-ray through vacuum ultraviolet energy range. A much more intense source is needed in this range. A possible solution was proposed which could provide not only circularly polarized photons over a wide energy range, but could in principle modulate the polarization of the beam between two orthogonal polarization states. Realization of this device, or an equivalent one, would be a vital step towards the goal of determining all components of the Mueller matrix for each spectroscopic experiment.

A variety of theoretical treatments are presented describing the different phenomena emerging from the interaction of matter and polarized radiation in a wide range of energies. From this work we expect to learn what are the most useful wavelength regions and what types of samples are the most suitable for study.

The hope of the editors and authors of this book is that readers will be stimulated to enter the field of polarized light spectroscopy. A wide range of problems is addressed here by current leaders of the field. Surely, better theories, more experiments on different systems, and completely new phenomena will be reported in the coming years. The field is young, and the opportunities are great.

Ignacio Tinoco, Jr.
Department of Chemistry
University of California

ACKNOWLEDGEMENTS

Organizing a conference is never an easy task. Bringing
scientists together from a variety of fields and interests to
exchange their experiences cannot be done merely by the good
intentions of the organizers. The participants themselves,
their willingness to interact, to discuss openly what they
understand or what they don't understand is what transforms a
gathering of scientists into an exciting learning experience.
In our case the ease and spontaneity with which the discussions
took place worked this magic. For us, as the organizers, this
was the greatest reward. To each of the participants we extend
our most sincere appreciation.

This conference would have not been made possible without
the financial support and help of the Departments of Chemistry
and Physics and the Office of the Associate Provost for Research
at the University of New Mexico. We also gratefully acknowledge
the support provided by the Center for High Technology Materials
Research.

Our special gratitude must go to Ms. Barbara Mosiello. She
not only dealt with the enormous amount of typing involved in
this endeavor, but with her typical enthusiasm she became
actively involved in the conference organization itself.

Finally, we want to thank all our graduate students who were
responsible for all local arrangements and an uncountable number
of small, but essential, details.

We hope that a scientific public wider than that able to
attend the conference will find in this volume a reference
source on the exciting field of the interaction of polarized
radiation and matter.

CONTENTS

CIRCULARLY POLARIZED VUV RADIATION
UP TO 40 eV PHOTON ENERGIES FOR
SPIN RESOLVED PHOTOELECTRON SPECTROSCOPY STUDIES

Ulrich Heinzmann

Fritz-Haber-Institute of the Max-Planck-Society
Faradayweg 4-6, D-1000 Berlin 33
West Germany

For many applications in spin resolved photoelectron spectroscopy circularly polarized radiation is required because the spin-polarization transfer from "spin polarized photons" onto spin polarized photoelectrons is the matter of interest. In atomic and molecular photoionization a lot of studies are hampered by the fact that most atoms and molecules have their ionization thresholds in the vuv, where conventional methods for producing circularly polarized radiation break down. There are different ways of obtaining circularly polarized vuv radiation.

(1) By using prisms, lenses or quarter wave plates built from material which is doubly refracting[1,2] in the vuv.
(2) By reflection on piles of plates or on metal surfaces at certain angles of incidence[3,4].
(3) By using circularly polarized vuv laser radiation[5,6,7].
(4) By use of the natural circular polarization of synchrotron radiation emitted out of the plane of the electron storage ring[8,9,10].
(5) By using the synchrotron radiation emitted by a helical wiggler of an electron storage ring[11,12].

While method (1) is restricted to a photon energy range smaller than about 10 eV because at higher energies no transparent material with birefringence exists, the use of the other four methods in the vuv range is only restricted by the specifications of the light source itself. Methods (2), (3) and (5) have only been proposed but not yet used to get circularly polarized vuv radiation so that the description here is restricted to (1) and (4); both have been used for spin resolved photoelectron spectroscopy studies.

Fig. 1 shows the set up of an apparatus[1] for the production of circularly polarized vuv radiation up to 9 eV. The radiation is produced by means of a Hinteregger lamp, whose H_2 many line spectrum is the most powerful "quasi continuum" in the wavelength range between 120 and 165 nm. The radiation monochromatized by a 0.5 m Seya Namioka vuv monochromator, linearly polarized by a MgF_2 Senarmont double prism and circularly polarized by a MgF_2 quarter wave double plate, intersects the atomic beam where the spin polarized photoelectrons are produced. They are extracted by an electric field and analyzed with respect to their spin polarization in a Mott detector[13]. The light is analyzed by a prism of the same type as used in the polarizer and is detected by a

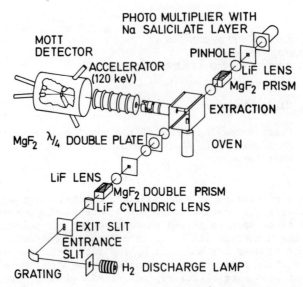

Fig. 1   Schematic diagram of the apparatus[1] to produce and use circu-
          larly polarized vuv radiation in the photon energy region up
          to 9 eV.

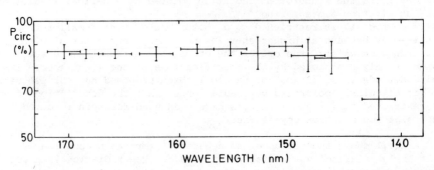

Fig. 2   Results[1] of degree of circular polarization measured as func-
          tion of the wavelength by use of the apparatus shown in Fig. 1.

photomultiplier with a sodium salicylate layer. By use of this apparatus measurements of the spin polarizations of photoelectrons ejected
by thallium[14,15], lead[13] and silver[16] atoms could be performed.

Opposite to standard prisms for linearly polarized light, for
example, Glan[17], in the arrangement in Fig. 1 both partial beams
of opposite linear polarization leave the prism at its front side; the
angle between the unrefracted (vertically polarized in Fig. 1) and the
refracted (horizontally polarized) beam is very small, between 0 and 3°
depending upon the wavelength which is a disadvantage of all prisms in
the vuv region. As also shown in Fig. 1 it is therefore necessary to
use a LiF lens before the prism in order to ensure the radiation is
parallel when it passes through the prism. The planocylindric lens in
front of the prism and the planoconvex lens behind it form a rectangu-
lar image of the monochromator exit slit at a position where a pinhole
separates the radiation beams of opposite linear polarization. Because
of the high dispersion of LiF in the vuv the lenses had to be moved
within distances of up to 10 cm, so that the exit slits and the pinhole
are always at the focus of the lenses. The circular polarizer consists
of two focussing (and moveable) LiF lenses and a MgF_2 quarter-wave
double plate: two plates have an angular difference of 90° between
their optic axes and a thickness difference such that the optical path
difference for the two linear polarization states is $\lambda/4$ at $\lambda = 150$ nm.

Fig. 2 shows the degree of circular polarization measured using
the arrangement shown in Fig. 1: it has been found to be about 85 to
90 % in the wavelength range 145 to 170 nm. The difference between the
measured values and the 100 %, especially the decline of the polarization to shorter wavelength at 140 nm is due to the impossibility to separate completely the two different linearly polarized radiation beams
by the pinholes unless the pinholes were to be partially closed cau-
sing a sharp reduction in the intensity of the transmitted radiation.
At wavelengths shorter than 140 nm the birefringence of MgF_2 decreases[18] and vanishes at 120 nm. Using the apparatus shown in Fig. 1
numbers of circularly polarized photons in the atomic beam region have
been obtained to be 4.10^{11} s^{-1} and 6.10^{10} s^{-1} at 158 nm and 142 nm,
respectively, with a bandwidth of 6 nm and a circular polarization
shown in Fig. 2. Radiation with half the bandwidth had a quarter of
this intensity[1].

According to Schwingers theory[19] synchrotron radiation is linearly
polarized in the plane of the electron beam; the radiation emitted into
directions above and below it is elliptically polarized with a high
fraction of circular polarization, as theoretically shown by Ref. 20
and 21 and as experimentally verified by Ref. 22 and by Ref. 8,23 in
the visible and in the vuv range, respectively. Two monochromators, a
10 m and a 6.3 m NIM, have been built for the circularly polarized
radiation emitted out of the plane, they are described in Ref. 8,23 and
9,10, respectively.

Fig. 3 shows the set up of the 6.5 m N.I. monochromator[9] which
is in operation at the new German dedicated electron storage ring BESSY.
It is of the Gillieson type, i.e. a spherical mirror which is adjustable
under UHV focus the radiation onto the exit slit, via a plane grating.
It is an advantage of this type of monochromator that no entrance slit
has to be used so that a wide horizontal angular range (54 mrad) limited by the horizontal apertures may be accepted to get highest intensities. The apertures movable in vertical direction allow the selection

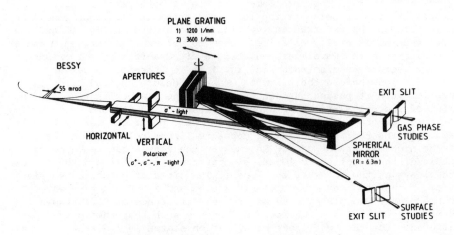

Fig. 3   Set-up of the Gillieson type monochromator[9] at BESSY for cir-
          cularly polarized synchrotron radiation.

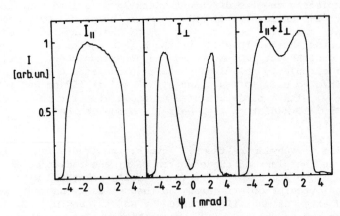

Fig. 4   Intensity dependence of the radiation components, polarized
          horizontally ($I_\parallel$) and vertically ($I_\perp$), upon the vertical
          angle ψ (± 0.25 mrad) the radiation is emitted with respect
          to the BESSY storage ring plane. $I_\parallel$ and $I_\perp$ and their sum have
          been measured[9] behind the monochromator exit slit at a wave-
          length of 450 nm.

of left handed or right handed circular polarization. Two different gratings interchangeable under vacuum are used to cover the wavelength
range from the visible to the vuv with different bandwidths. The wavelength scan is performed by a simultaneous rotation and, to maintain
focusing, an independent translation of the grating; both as well as
the motion of the vertical apertures are controlled by a computer. The
UHV monochromator has two exit slits, which are used for studies of
atomic and molecular photoionization and for surface and bulk photoemission studies, respectively. The bandwidth of the radiation behind the
exit slits depends upon the dispersion of the monochromator, which is
5.4 mm/nm and 16.2 mm/nm for the 1200 and 3600 lines mm^{-1} grating, respectively, and upon the widths of the electron beam in the storage ring
and of the exit slit. A bandwidth of 0.5 nm for the use of the 1200 l/mm
grating is obtained during typical operation of BESSY.

Fig. 4 shows the components $I_{\parallel}$ and $I_{\perp}$ of the radiation emitted,
whose electric field vector oscillates parallel and perpendicular with
respect to the orbital plane of the storage ring, and the total intensity[9]. They are function of the vertical angle ψ the radiation is
emitted with respect to the orbital plane of BESSY ($\psi = 0$). The shapes
and the halfwidths of the profiles are in good agreement with the expected calculated values although the actual position of the orbital
plane which may change slightly after each injection seems to be a
little bit too low. In the plane ($\psi = 0$) $I_{\perp}$ vanishes within the experimental uncertainty, i.e. there the radiation emitted is linearly polarized in the plane.

Fig. 5 shows the degree of linear and circular polarization P_{lin}
and P_{circ} as function of the vertical angle ψ measured[10] by use of a
rotatable arrangement of four gold coated mirrors[8]. Even at short wavelength of 50 nm very high degrees of linear and circular polarization,
closely to 100 %, have been obtained. The radiation emitted above the
plane is right handed, the radiation emitted below the plane is left
handed circularly polarized. The horizontal error bars describe the
slit width of the vertical aperture. Using the slit width shown corresponding to ± 0.1 mrad the intensity, however, was not very high. Therefore in most applications very large slitwidths have been used due to
intensity reasons. As reported by Ref. 24 more than 10^{11} circularly polarized photons s^{-1} have been obtained in the vuv range which enabled
angular, energy and spin resolved photoionisation and photoemission studies with circularly polarized radiation on atoms[10], solids[24] and adsorbates[25], for the first time. Fig. 6 shows the degrees of linear and circular polarization in the angle integrated case where the lower edge of
the aperture is at ψ and the upper edge is at ± 5 mrad. It shows that
for $\psi = 0$ i.e. the upper half of the synchrotron radiation is accepted
and analyzed, the overall averaged circular polarization has been found
to be 85 % at the wavelength of 100 nm. The lower half of the synchrotron radiation has then a degree of circular polarization of - 85 %.
Under these conditions the intensity loss to get circularly polarized
radiation compared with the case all synchrotron radiation is accepted
is only a factor of two. If the apertures are open from ± 1 mrad to
± 5 mrad a photon flux on the order of 10^{11} s^{-1} (200 mA beam current in
the storage ring) of a degree of circular polarization of 93 % passed
the monochromator exit slit, which corresponds with an intensity loss
factor of five.

In Fig. 7 the wavelength dependence of the circular polarization is
shown in the vuv range. The results corresponding to the data in Fig. 6

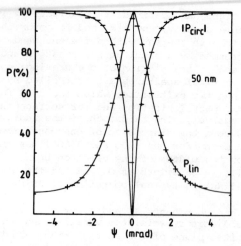

Fig. 5   Dependence[10] of the circular and linear polarization of the
          vuv radiation on the vertical angle ψ the radiation is emit-
          ted out of the BESSY plane and analyzed behind the exit slit
          of the 6.3 m N.I. monochromator.

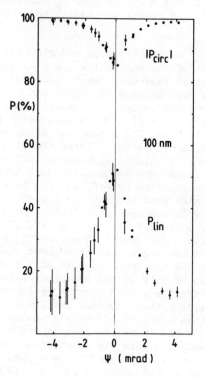

Fig. 6   the same as in Fig. 5 but as function of the vertical angular
          range[10] accepted from ψ to ± 5 mrad.

the lower error bars for an open vertical angular range from 0 to 3.5 mrad and the upper error bars for the vertical angular range from 1 mrad to 3.5 mrad, have been, however, obtained[8] at the 10 m N.I. monochromator, shown in Fig. 8, at the synchrotron of the university Bonn. The circular polarization is wavelength independent within the experimental uncertainty in the wavelength range between 40 and 100 nm. Furthermore the results in Fig. 7 differ from those in Fig. 6 in that the degree of circular polarization of the radiation emitted by the Bonn snychrotron is about 10 % lower than in the case the radiation is emitted by the storage ring BESSY. This deviation which agrees quantitatively with the deviation between experimental results and the calculated values according to Schwingers theory as discussed by Ref. 8 is due to the fact that vertical synchrotron beam oscillations play a more important role at synchrotrons than at electron storage rings.

It is worth noting that the optical components have a negligible influence on the light polarization as the measured results show, because the monochromators work with normal incidence reflections and Littrow mounting at mirrors and gratings, respectively. Last not least the measurements of Ref. 8 have shown that synchrotron radiation emitted into a certain direction out of the plane of the storage ring is completely elliptically polarized, i.e. that no unpolarized background emitted by the electron beam and monochromatized in the monochromator has to be taken into account in the studies behind the exit slits.

Using the circularly polarized synchrotron radiation photoelectron emission studies with atoms[10,23,26-30], molecules[31-33], adsorbates[25] and solid states[24] could be performed, which have been in some cases simultaneously resolved with respect to the radiation wavelength, the radiation polarization, the photoelectron emission angle, the kinetic energy of the photoelectron as well as the components of the spin polarization vector of the photoelectrons. As examples Fig. 9 and Fig. 10 show the angular dependence of the spin polarization and the wavelength dependence of the angle integrated spin polarization of photoelectrons ejected from xenon and krypton atoms by circularly polarized radiation, respectively. Very high electron polarization values between - 60 % and + 100 % have been found. A detailed description and analysis of these data especially also in the comparison between the theoretical predictions (partly shown in the figures) and the experimental results is given by Ref. 10 and 29 and by the references therein.

The use of circularly polarized synchrotron radiation emitted out of the plane of the storage ring opened a new field of photoelectron spectroscopy: it has been found rather common than exceptional that photoelectrons are polarized. Using this new information quantitatively, a complete quantum mechanical characterization of the photoionization process in terms of experimentally determined matrix elements and phase shift differences of wave functions could be performed for some atoms[26,28,29]. It is planned to be continued for molecules and adsorbates but even also for more complicated systems like solid state surfaces.

The author wishes to express his gratitude to his coworkers Drs. A. Eyers, Ch. Heckenkamp, F. Schäfers and G. Schönhense for their encouragement and acknowledges support by the BMFT.

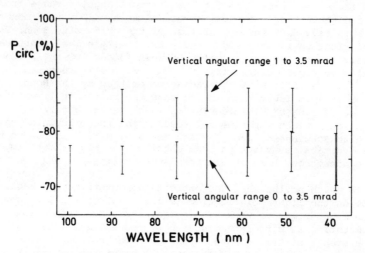

Fig. 7   Wavelength dependence of the degree of circular polariza-
            tion[8,23] of the synchrotron radiation emitted out of the plane
            of the synchrotron in Bonn within the vertical angular ranges
            from 0 to 3.5 mrad and from 1 to 3.5 mrad monochromatized by
            10 m N.I. monochromator shown in Fig. 8.

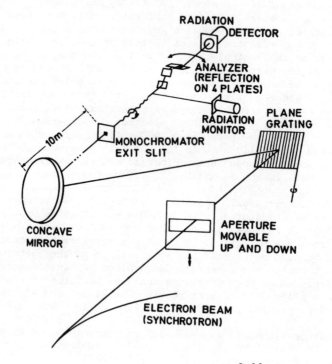

Fig. 8   Set-up of the 10 m N.I. monochromator[8,23] at the 2.5 GeV syn-
            chrotron in Bonn.

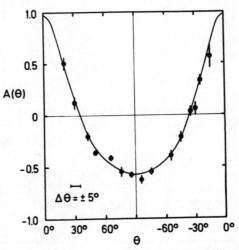

Fig. 9   Angular dependence of the spin-polarization component paral-
         lel to the spin of the incoming photons (80 nm) for photo-
         electrons leaving the xenon ion in the $^2P_{1/2}$ state; the full
         line is a fit to the experimental points (Ref. 10).

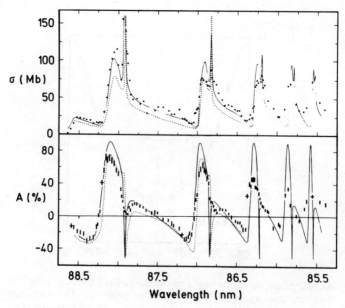

Fig. 10   Emission angle integrated photoionization cross section (upper half) and photoelectron spinpolarization (lower half) in
the autoionization range of krypton atoms[29].

REFERENCES

1. U. Heinzmann, J. Phys. E 10, 1005 (1977)
2. P.A. Snyder and E.M. Rowe, Nucl. Instrum. Meth. 172, 345 (1980)
3. M. Schledermann and M. Skibowski, Appl. Opt. 10, 321 (1971)
4. P.D. Johnson and N.V. Smith, Nucl. Instrum. Meth., in press (1984)
5. C.R. Vidal, invited talk at the "Workshop on Techniques for the Production and Utilization of Polarized Radiation in the VUV", Imperial College London 1982, unpublished
6. R. Wallenstein, Optics Commun. 33, 119 (1980)
7. H. Zacharias, H. Rottke and K.H. Welge, Optics Commun. 35, 185 (1980)
8. U. Heinzmann, B. Osterheld and F. Schäfers, Nucl. Instrum. Meth. 195, 395 (1982)
9. A. Eyers, Ch. Heckenkamp, F. Schäfers, G. Schönhense and U. Heinzmann, Nucl. Instrum. Meth. 208, 303 (1983)
10. Ch. Heckenkamp, F. Schäfers, G. Schönhense and U. Heinzmann, Phys. Rev. Lett. 52, 421 (1984)
11. E.L. Salzin and Yu. M. Shatunov, XI. USSR Conference on accelerators, Dubna 1978, Proceedings Vol. 1, p. 124 and G.A. Kezerashvili, P.P. Lysenko, V.M. Khorev, G.M. Tchernykh and Yu. M. Shatunov, Inst. of nuclear physics Novosibirsk, USSR, private communication 1982
12. H. Winnick, H. Wiedemann, I. Lindau, K. Hodgson, K. Halbach, J. Cerino, A. Bienenstock and R. Bachrach, IEEE Trans. Nucl. Sci., in press (1984)
13. U. Heinzmann, J. Phys. B 11, 399 (1978)
14. U. Heinzmann, H. Heuer and J. Kessler, Phys. Rev. Lett. 34, 441 (1975)
15. U. Heinzmann, H. Heuer and J. Kessler, Phys. Rev. Lett. 36, 1444 (1976)
16. U. Heinzmann, A. Wolcke and J. Kessler, J. Phys. B. 13, 3149 (1980)
17. R. Möllenkamp and U. Heinzmann, J. Phys. E 15, 692 (1982)
18. V. Chandrasekharan and H. Damany, Appl. Opt. 8, 671 (1969)
19. J. Schwinger, Phys. Rev. 75, 1912 (1949)
20. H. Olsen, Kgl. Norske Vidensk. Selsk. Skrifter, No 5 (1952)
21. K.C. Westfold, Astrophys. J 130, 231 (1959)
22. P. Joos, Phys. Rev. Lett. 4, 558 (1960)
23. U. Heinzmann, J. Phys. B 13, 4353 (1980)
24. A. Eyers, F. Schäfers, G. Schönhense, U. Heinzmann, H.P. Oepen, K. Hünlich, J. Kirschner and G. Borstel, Phys. Rev. Lett. 52, 1559 (1984)
25. G. Schönhense, A. Eyers, U. Friess, F. Schäfers and U. Heinzmann, to be published, 1984
26. U. Heinzmann, J. Phys. B 13, 4367 (1980)
27. U. Heinzmann and F. Schäfers, J. Phys. B 13, L 415 (1980)
28. F. Schäfers, G. Schönhense and U. Heinzmann, Z. Phys. A 304, 41 (1982)
29. F. Schäfers, G. Schönhense and U. Heinzmann, Phys. Rev. A 28, 802 (1983)
30. Ch. Heckenkamp, F. Schäfers, U. Heinzmann, R. Frey and E.W. Schlag, Nucl. Instrum. Meth. 208, 805 (1983)
31. U. Heinzmann, F. Schäfers and B.A. Hess, Chem. Phys. Lett. 69, 284 (1980)
32. U. Heinzmann, B. Osterheld, F. Schäfers and G. Schönhense, J. Phys. B 14, L 79 (1981)
33. F. Schäfers, M.A. Baig and U. Heinzmann, J. Phys. B 16, L 1 (1983)

CIDS CALCULATIONS ON QUARTZ AT HARD X-RAY WAVELENGTHS

David Keller and Carlos Bustamante

Department of Chemistry
University of New Mexico
Albuquerque, NM 87131

1. INTRODUCTION

When the prospect of performing optical activity experiments
at short wavelengths is discussed, it is natural to ask how large
optical activity effects will be at short wavelengths, and what
is the mechanism by which these effects arise? As a first step
toward answering these questions, we report here an attempt to
calculate the circular intensity differential scattering (CIDS)
of α-quartz, an optically active crystal with a helical unit
cell. The calculation was ultimately not quantitatively
successful, and a somewhat different approach is perhaps
necessary. But despite these shortcomings some useful conclusions
can be drawn, especially concerning the mechanism by which
optical activity arises at X-ray wavelengths.

The CIDS is defined as the ratio of the difference in
scattered intensities when left- and right-circularly polarized
light are incident on the sample, divided by their sum:

$$\text{CIDS} = \frac{I_L(\theta,\phi) - I_R(\theta,\phi)}{I_L(\theta,\phi) + I_R(\theta,\phi)}$$

where $I_{L,R}(\theta,\phi)$ is the light intensity scattered in the
direction given by the angles θ and ϕ with respect to the
incident beam, for incident left (L) or right (R) circularly
polarized light. More is said about the general properties of
CIDS and the results of experimental measurements in the papers
by K. Hall et al., M. F. Maestre et al., and C. Nicolini. Also,
in the papers by W. Bickel and M. McClain on the Mueller matrix,

the CIDS is included as a special case. (The CIDS is the S_{14} element of the Mueller matrix.) Finally, if the system has an appreciable magnetic susceptibility, there can be contributions to the CIDS from magnetic scattering, which we do not consider here.

2. THEORY

Our calculations are based on the scattering theory described by Saxon[1], and we will give only a brief derivation of the central results here. The theory is a classical one, and so the derivation begins with the Maxwell equations for a system containing no free charges or currents. We have also neglected any magnetic susceptibility.

a) $\nabla \cdot D = 0$,   b) $\nabla \cdot B = 0$

c) $\nabla \times E + \frac{1}{c}\frac{\partial B}{\partial t} = 0$,   d) $\nabla \times B - \frac{1}{c}\frac{\partial D}{\partial t} = 0$ (1)

As our constitutive equation connecting the electric displacement D with the electric field E, we have

$$D(x,t) = \varepsilon(x) \cdot E(x,t) \qquad (2)$$

where $\varepsilon(x)$ is a spatially varying dielectric tensor. Physically, it is the spatial gradients in $\varepsilon(x)$ which give rise to scattering. We also take E and B to have harmonic time dependence

$$E(x,t) = e^{-i\omega t}E(x)$$
$$B(x,t) = e^{-i\omega t}B(x) \qquad (3)$$

where ω is the frequency of oscillation of the fields. This step amounts to ignoring any frequency shifts that occur during the scattering process. We are therefore allowing only for _elastic_ scattering and neglect the effects of inelastic scattering.

With equations (2) and (3) it is possible to manipulate the Maxwell equations (1) into a wave equation for E. From equation (1c)

$$\nabla \times E - ikB = 0$$

where $k = \omega/c$. Taking the curl, we obtain

$$\nabla \times (\nabla \times E) - ik\nabla \times B = 0.$$

From equation (1d)

$$\nabla \times \underset{\sim}{B} = -ik\underset{=}{\varepsilon}(\underset{\sim}{x}) \cdot \underset{\sim}{E},$$

so

$$\nabla \times (\nabla \times \underset{\sim}{E}) - k^2 \underset{=}{\varepsilon}(\underset{\sim}{x}) \cdot \underset{\sim}{E} = 0. \tag{4}$$

We can convert equation (4) into an integral equation by use of the tensor Green's function, $\underset{=}{\Gamma}(\underset{\sim}{x},\underset{\sim}{x}')$. Using the properties of a Green's function, and after some manipulation, we obtain

$$\underset{\sim}{E}(\underset{\sim}{x}) = \underset{\sim}{E}_o(\underset{\sim}{x}) + 4\pi k^2 \; \underset{=}{\Gamma}(\underset{\sim}{x},\underset{\sim}{x}') \cdot \frac{\underset{=}{\varepsilon}(\underset{\sim}{x})-\underset{=}{1}}{4\pi} \cdot \underset{\sim}{E}(\underset{\sim}{x}') d^3 x' \tag{5}$$

where $\underset{=}{\Gamma}(\underset{\sim}{x},\underset{\sim}{x}')$ satisfies

$$\nabla \times (\nabla \times \underset{=}{\Gamma}(\underset{\sim}{x},\underset{\sim}{x}')) - k^2 \underset{=}{\Gamma}(\underset{\sim}{x},\underset{\sim}{x}') = \underset{=}{1}\delta^3(\underset{\sim}{x}-\underset{\sim}{x}'). \tag{6}$$

$\underset{=}{\Gamma}(\underset{\sim}{x},\underset{\sim}{x}')$ is given explicitly by

$$\underset{=}{\Gamma}(\underset{\sim}{x},\underset{\sim}{x}') = (\underset{=}{1} - \frac{1}{k^2} \nabla\nabla') \frac{e^{ik|\underset{\sim}{x}-\underset{\sim}{x}'|}}{4\pi|\underset{\sim}{x}-\underset{\sim}{x}'|}$$

$$= (\underset{=}{1}-\hat{\underset{\sim}{r}}\hat{\underset{\sim}{r}}) + (3\hat{\underset{\sim}{r}}\hat{\underset{\sim}{r}}-\underset{=}{1}) \frac{1}{k^2 r^2} - \frac{i}{kr} \frac{e^{ikr}}{4\pi r} - \frac{\underset{=}{1}}{3k^2} \delta^3(\underset{\sim}{r}) \tag{7}$$

where $r = x-x'$, $r = |\underset{\sim}{x}-\underset{\sim}{x}'|$, and $\hat{\underset{\sim}{r}} = \underset{\sim}{r}/r$. The incident electric field $\underset{\sim}{\tilde{E}}_o(\underset{\sim}{x})$ must satisfy

$$\nabla \times (\nabla \times \underset{\sim}{E}_o) - k^2 \underset{\sim}{E}_o = 0.$$

The usual choice for $\underset{\sim}{E}_o$ is a transversal plane wave,

$$\underset{\sim}{E}_o = E_o \hat{\underset{\sim}{\varepsilon}}_o e^{i\underset{\sim}{k}_o \cdot \underset{\sim}{x}},$$

where E_o, $\hat{\underset{\sim}{\varepsilon}}_o$, and $\hat{\underset{\sim}{k}}_o$ are the amplitude, polarization, and wavevector of the incident light, respectively

In equation (5) all boundary conditions are automatically accounted for. Also the incident field, $\underset{\sim}{E}_o$ now appears explicitly in the equation we have to solve.

For the case of α-quartz at X-ray wavelengths, we use a microscopic form of equation (5), with the atoms of the quartz crystal represented by discrete point scatterers. This amounts to the following replacement:

$$\underset{\approx\sim}{\chi}(x) \equiv \frac{\underset{\approx}{\varepsilon}(x) - \underset{\approx}{1}}{4\pi} \longrightarrow \underset{\approx\sim}{\alpha}(x)$$

where $\underset{\approx}{\alpha}(x)$ is the microscopic polarizability density. For discrete atoms $\underset{\approx}{\alpha}$ is given by

$$\underset{\approx\sim}{\alpha}(x) = \sum_{i=1}^{N} \underset{\approx i}{\alpha}\, \delta^3(\underset{\sim}{x}-\underset{\sim i}{x})$$

where $\underset{\approx i}{\alpha}$ is the polarizability tensor associated with the ith atom in the crystal, x_i is the position of the ith atom, and N is the number of atoms in the system. With these substitutions equation (5) becomes

$$\underset{\sim}{E}(\underset{\sim}{x}) = \underset{\sim o}{E}(\underset{\sim}{x}) + 4\pi k^2 \sum_{i=1}^{N} \underset{\approx\sim\sim i}{\Gamma}(x,x_i) \bullet \underset{\approx i}{\alpha} \bullet \underset{\sim}{E}(\underset{\sim i}{x}) \tag{8}$$

Though equation (8) was arrived at by completely formal means, it has a simple physical interpretation. The quantity $\underset{\approx i}{\alpha} \bullet \underset{\sim}{E}(\underset{\sim i}{x})$ appearing in equation (8) is just the electric dipole moment μ_i induced in the ith atom by the electric field $\underset{\sim}{E}(\underset{\sim i}{x})$. It can also be shown (either by direct calculation or by appeal to the physical interpretation of a Green's function) that the quantity $\underset{\sim i}{E} \equiv 4\pi k^2\, \underset{\approx}{\Gamma}(x,x_i) \bullet \mu_i$ is just the electric field produced by the oscillating dipole moment $\underset{\sim i}{\mu}$. Equation (8) is therefore a statement that the total electric field $E(x)$ is the superposition of the incident electric field $\underset{\sim}{E}_o(x)$ and the sum of all the fields produced by the oscillating dipole moments induced in the system. The <u>scattered</u> electric field is then just the sum of the dipole fields (i.e., the total field $\underset{\sim}{E}(\underset{\sim}{x})$ minus the incident field).

The difficulty in solving equation (8) lies in finding the quantitites $E(\underset{\sim i}{x})$, which are the values of the electric field at each atom inside the crystal. These fields are due both to the incident field $\underset{\sim}{E}_o$ and to the fields produced by the oscillating dipole moments of the atoms. In principle we could use equation (8) to set up a system of 3N equations in 3N unknowns and solve for the exact $E(\underset{\sim i}{x})$'s. This approach is not practical for a system with a large number of scatterers however. Instead we have chosen to use an iterative procedure. We first make the approximation that the internal electric field is just the incident electric field. With this approximation we calculate a new approximate internal field. Then we use the new field to calculate a still better approximation to the internal field, and so on. If the polarizabilities of the scatterers are not too large, the process should converge fairly rapidly. In practice we found that the total scattered intensity does converge rapidly for α-quartz, but the CIDS, or differential scattering, does not. More will be said about this later.

The crystal structure of α-quartz can be described as a system of parallel, interdigitated helices with three Si_2O_2 "molecules" per turn of helix. The helix pitch is 5.40Å, and the radius is 2.31Å.[2] One turn of one helix may be chosen as the unit cell for the crystal (3 Si_2O_2 molecules, 9 atoms). For our calculations each atom was assigned an isotropic polarizability, with polarizability magnitude calculated from tables of atomic scattering factor data according to the relation

$$\alpha = \frac{\lambda^2 e^2}{4\pi^2 mc^2} (f_1 + if_2) \tag{9}$$

where f_1 and f_2 are the real and imaginary parts of the scattering factor, f.[3] The constant factor $e^2/(4\pi^2 mc^2)$ is equal to 7.13×10^{-15} cm in Gaussian units. At $\lambda = 1.54$Å the polarizabilities for oxygen and silicon are:[3]

$$\alpha_{oxy} = 1.35 \times 10^{-5} + i\, 5.70 \times 10^{-6} \text{ Å}^3$$

$$\alpha_{sil} = 2.03 \times 10^{-5} + i\, 5.41 \times 10^{-7} \text{ Å}^3.$$

In all of our calculations we took the incident radiation to be along the helical axis of the quartz crystal and evaluated the CIDS in only one scattering plane.

In doing our calculation we cannot, of course, use an infinite crystal. The iterative procedure described above allows us to use a relatively large number of scattering atoms in our calculation, but even so we are limited to at most a few hundred. This means that the "crystal" we can model must be quite small (<50 unit cells). This limitation is not as serious for CIDS calculations as it would be for normal scattering calculations however, because in taking the CIDS ratio the primary effects of the lattice periodicity divide out, and what remains depends mostly on the properties of a single unit cell. To see how this happens we note first that it can be shown that the scattered electric field for a periodic system is a product of two factors:[4]

$$\underset{\sim}{E}_{scatt}(\underset{\sim}{x}) = F(\underset{\sim}{k},\underset{\sim}{k}_o)\, \underset{\sim}{A}\,(\underset{\sim}{x})$$

where

$$F = \sum_{pqr} e^{-i(\underset{\sim}{k}-\underset{\sim}{k}_o)\bullet(p\underset{\sim}{a}+q\underset{\sim}{b}+r\underset{\sim}{c})},$$

and where $\underset{\sim}{k}$ is the wavevector of the scattered radiation in the direction of observation, $\underset{\sim}{a}$, $\underset{\sim}{b}$, $\underset{\sim}{c}$ are the unit cell vectors of the crystalline lattice, and $\underset{\sim}{p}$, $\underset{\sim}{q}$, $\underset{\sim}{r}$, are integers, and

$$\underset{\sim}{A} = k^2\frac{e^{ikr}}{r}(1-\hat{k}\hat{k})\bullet\sum_{\underset{\sim}{i}\ \text{unit cell}} e^{-i(\underset{\sim}{k}-\underset{\sim}{k}_o)\bullet\underset{\sim}{x}_i}\ \underset{=i}{\alpha}\bullet\underset{\sim}{E}(\underset{\sim}{x}_i) \qquad (10)$$

where r is the distance from the sample to the detector, and the sum is only over the atoms of one unit cell. The quantity A is not independent of the fact that the unit cell is part of a larger crystal because the quantity $E(\underset{\sim}{x}_i)$, the internal field, is affected by atoms outside the unit cell as well as those inside. The quantity F is the lattice factor familiar in the theory of X-ray diffraction and causes the scattered radiation to be "quantized" into discrete directions in space. It is this "quantization" of the scattered radiation that disappears in taking the CIDS ratio. Since F is independent of the polarization of the incident radiation, the scattered intensities for left (L) or right (R) polarized light are given by:

$$I_{L,R} \propto |F|^2\ |\underset{\sim}{A}_{L,R}|^2,$$

and so,

$$CIDS = \frac{I_L-I_R}{I_L+I_R} = \frac{|\underset{\sim}{A}_L|^2-|\underset{\sim}{A}_R|^2}{|\underset{\sim}{A}_L|^2+|\underset{\sim}{A}_R|^2}. \qquad (11)$$

In most of our calculations we used a 3x3x3 "cube" of 27 unit cells (243 atoms) as our model. This system was chosen because it provides the unit cell in the middle of the cube with a layer of unit cells on all sides to give an estimate of the correct internal field. We first calculated the internal field using 27 unit cells, and then in a separate step we calculated the scattered field (and the CIDS ratio) using only the atoms of the central unit cell as scatterers, but using the internal field that was found using the 27 unit cells. According to equations (10) and (11) this procedure gives the correct CIDS ratio provided our estimate of the internal electric field is accurate.

Two problems emerged in the actual calculations. First, though we went as high as the fourth cycle in the iterative procedure ("fourth Born approximation"), the CIDS did not appear to be converging. Second, when we added extra unit cells to our original 3x3x3 cube, we found that the calculated CIDS was not the same as before, indicating that our system was not large enough to estimate the internal field to the necessary accuracy.

It became apparent in the course of further calculations that the cause of our troubles was the great sensitivity of the CIDS for systems of isotropic scatterers to the details of the internal field. It can be shown analytically that in the first Born approximation (i.e., when the internal field is taken to be the incident field alone) there can be no CIDS unless the scatterers are anisotropic, even though they may be arranged in a chiral fashion.[4] When better approximations to the internal field are made, it is found that chiral systems of isotropic scatterers can show CIDS, but it is clear that these CIDS effects will depend strongly on the details of the induced interactions between scatterers (i.e., the fields produced by other scatterers in the system), since it is these interactions that give rise to the corrections to the first Born approximation. This situation is similar to what happens in circular dichroism, where once again the optical activity effect depends on induced couplings between adjacent subunits. In calculations where we added 1% anistropy to the atom polarizabilities the calculations converged rapidly and were insensitive to extra unit cells.

In the calculations we found values in the range 10^{-2} to 10^{-5}. The larger values were found only when we made the polarizabilities anisotropic. No experiments have been done to establish what degree of anisotropy may exist in quartz, but it seems likely to be quite small ($<10^{-3}$ at 1.54Å)[5].

An important aspect of the interaction of radiation with crystalline systems is the tendency for long-range induced waves to be established inside the crystal. These waves determine the Bragg diffraction directions and are the basis for the dynamical theory of x-ray diffraction. It seems likely that the dynamic coupling of one scattering atom to the remainder of the crystal, which will give rise to optical activity, is also mediated primarily through such waves. It is, however, precisely these long-range couplings which we have excluded in taking a small crystallite as representative of the entire crystal. For this reason it seems to us that our calculation can be considered no better than a lower bound for the CIDS in a very imperfect crystal.

In conclusion then, though we found that CIDS magnitudes for α-quartz are quite small, these results must be taken very provisionally until better theoretical treatments are done. An approach based on the dynamical theory of x-ray diffraction mentioned above seems the most promising at this point.

ACKNOWLEDGEMENTS

This work was supported in part by National Institutes of Health grant GM 32543 (C.B.), 1984 Searle Scholarship granted to C.B., a grant by Research Corporation (C.B.), and a grant from Sandia Laboratories (C.B.).

REFERENCES

1. Saxon, D. S., "Lectures on the Scattering of Light", Sci. Rep., No. 9, Contract AF 19(122)-239, Dept. of Meteorology, University of California, Los Angeles (1955).

2. Zachariasen, W. H., and H. A. Plettinger, Acta Cryst., 18, 710-714 (1965).

3. Henke, B. L. , Lee, P., Tanaka, T. J., Shimabukuro, R. L., and B. K. Fujikawa, Atomic Data and Nuclear Data Tables, 27, 1-144 (1982).

4. Bustamante, C., Maestre, M. F., Keller, D., and I. Tinoco, Jr., J. Chem. Phys., 80, 4817-4823 (1984).

5. Templeton, D., personal communication (1984).

A CROSSED UNDULATOR SYSTEM FOR A VARIABLE POLARIZATION

SYNCHROTRON RADIATION SOURCE

Kwang-Je Kim

Lawrence Berkeley Laboratory

One Cyclotron Road, Berkeley, CA 94720

Abstract: A crossed undulator system can produce synchrotron
radiation whose polarization is arbitrary and adjustable. The
polarization can be linear and modulated between two mutually
perpendicular directions, or it can be circular and can be modulated
between right and left circular polarizations. The system works on
low emittance electron storage rings and can cover a wide spectral
range. Topics discussed include the basic principle of the system,
the design equations and the limitations in performance.

(I) Introduction

Magnetic structures in modern storage rings are intense sources of
radiation[1], called synchrotron radiation, which covers a wide
region of the photon energy spectrum. However, the applications of
synchrotron radiation to polarization sensitive experiments have been
limited because the polarization state of the synchrotron radiation
could not be changed readily. The crossed undulator system [2]

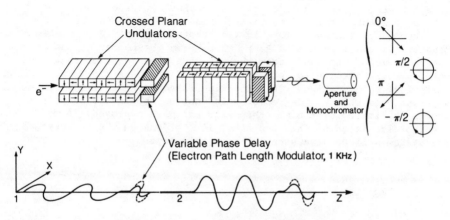

Fig. (1) Operating Principle of the Crossed Undulator System

21

discussed here removes such limitations and offers the possibility of complete polarization control. Thus, the polarization can be rapidly modulated between, for example, left circular and right circular.

After describing the basic principle of the operation in Section (II), various formulas useful in designing the system are derived in Section (III). In Section (IV) the performance of the system is studied, taking into account the electron beam angular spread and also the finite bandpass of the monochromator. Section (V) gives an example system design based on the VUV ring at NSLS. Finally, Section (VI) contains some concluding remarks.

(II)   Basic Principle

The operation of the crossed undulator system is illustrated in Fig. (1). An electron oscillates in the x-direction as it passes through the first undulator and radiates photons linearly polarized in the x-direction. In passing through the second undulator, the oscillation is along the y-direction with the resulting radiation polarized along the same direction. The radiations, when observed through a monochromator, combine to give rise to an elliptical polarization vector

$$\underset{\sim}{e} = \frac{1}{\sqrt{2}} \, (\hat{x} + e^{i\emptyset}\hat{y}) \tag{1}$$

Here $\hat{x}$ and $\hat{y}$ are unit vectors along the x and y directions, respectively, and $\emptyset$ is the relative phase between radiations from the two undulators. The phase $\emptyset$ is controlled by a variable field magnet called the modulator (shaded block in Fig. (1), which introduces electron path length modulation. By a proper adjustment of $\emptyset$, it is possible to obtain linear, circular or a general elliptical polarization.

The crossed undulator system produces two wave trains, one from the first undulator polarized along the x-direction, the other from the second undulator polarized along the y-direction. These two wave trains are separated in space. For our purpose, these waves need to interfere to produce an elliptically polarized wave. The superposition of the two waves can be achieved by the action of a monochromator[3], which is to stretch a short pulse of radiation into a long wave train of n = λ/δλ periods, where δλ is the monochromator bandpass. This is illustrated in Fig. (2.a). When the bandpass is small so that n >> N, where N is the number of undulator magnet periods, the waves from the two undulators overlap each other after passing through the monochromator and become a single wave of an elliptical polarization. This is illustrated in Fig. (2.b). In this way, one sees that the degree of polarization obtainable in this scheme is limited by the monochromator bandpass employed. A more precise discussion of this and other limitations on the performance of the system will be discussed later in this paper.

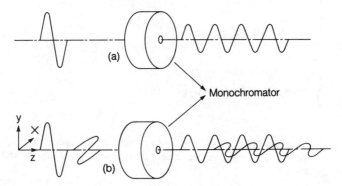

Fig. (2) Monochromater Action

(III) Analysis and Design of a Crossed Undulator System

A. General Analysis

From standard electromagnetic theory[4], the number of photons radiated into the forward direction per unit solid angle by an electron during its motion through a crossed undulator system is given by

$$\frac{dN}{d\Omega} = \alpha \frac{\delta\lambda}{\lambda} |\underset{\sim}{\varepsilon}|^2 \ , \tag{2}$$

$$\underset{\sim}{\varepsilon} = \frac{1}{\lambda} \int_{z_1}^{z_3} dz \ \underset{\sim}{\beta} (z) e^{ik\int_{z'}^{z} (1 - \beta_z(z'))dz'} \tag{3}$$

Here α is the fine structure constant ($= 1/137$), $\delta\lambda$ is the bandpass, λ is the photon wavelength, $k = 2\pi/\lambda$, and

$$\underset{\sim}{\beta} = (\underset{\sim}{\beta}, \beta_z) = (\beta_x, \beta_y, \beta_z)$$

is the electron's velocity divided by the velocity of light. The coordinate system here is the one shown in Fig. (1). The limits of integration z_1 and z_3 in Eq (3) are the z coordinates of the entrance to the first undulator and the exit of the second undulator, respectively.

The second undulator is assumed to be identical to the first except that its orientation is rotated by 90° relative to the first one. The radiation vector ε can then be expressed in terms of those quantities involving the first undulator only as follows:

$$\underset{\sim}{\varepsilon} = (\hat{x} + e^{i\emptyset} \hat{y}) \ \varepsilon_o. \tag{4}$$

$$\emptyset = \frac{2\pi}{\lambda} \int_{z_1}^{z_2} dz \ (1 - \beta_z(z)), \tag{5}$$

$$\varepsilon_0 = \frac{1}{\lambda} \int_{z_1}^{z_2} dz \ \beta_x(z) e^{ik \int_{z_1}^{z} (1-\beta_z(z'))dz'}, \tag{6}$$

Here z_2 is the z-coordinate of the entrance of the second undulator. $\emptyset$ in Eq (4) specifies the polarization ellipse and was introduced in Eq (1).

B.   Design Objectives

For a satisfactory polarization control, the crossed undulator system needs to satisfy the following requirements:

i)   The radiation spectrum has a peak at the desired photon energy.
ii)   The polarization can be modulated rapidly between two arbitrary states, e.g., the left circular and the right circular polarizations.

iii)   The radiation intensity remains unchanged as the polarization modulates.

Corresponding to these three capabilities, each undulator of the crossed undulator system will be assumed to consist of three distinct magnetic parts. The first part serves the function (i) in the above, and controls the overall radiation intensity. It will be characterized by a subscript u, and has N periods with period length λ_u. The second part, called the corrector, serves the function

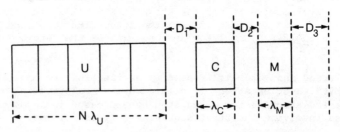

Fig. (3) Three Parts of Undulator

(iii) and is a single period magnet of length λ_c. The third part, called the modulator, serves the function (ii) and is a single period magnet with period length λ_m. These are shown in Fig. (3), in which the distances D_1, D_2, and D_3 are also specified. Here D_3 is the distance between the exit of the modulator to the entrance of the second undulator.

The magnetic field within each part will be assumed to be sinusoidal with peak field B_i (i = u, c, m). One also defines the magnet strength parameter K_i via

$$K_i = .934\ \lambda_i\ (cm)\ B_i\ (Tesla). \tag{7}$$

The parameter K is useful in characterizing the electron trajectory in undulator fields.[5]

C.   Derivation

With the above assumptions, electron motion in the undulator is completely specified and one obtains

$$\emptyset = \frac{\pi}{\lambda\gamma^2}\ (N\lambda_u(1 + K_u^2/2) + \lambda_c\ (1 + K_c^2/2) + \lambda_m(1 + K_m^2/2) + D) \tag{8}$$

Here $\S$ is electron energy in units of its rest energy and $D = D_1 + D_2 + D_3$. We assume N >> 1, in which case the undulator spectrum is peaked at the wave length

$$\lambda_1 = \frac{\lambda_u(1 + K_u^2/2)}{2\gamma^2} \tag{9}$$

At this wave length, Eq (8) becomes

$$\emptyset = 2\pi\ \left(N + \frac{\lambda_c(1 + K_c^2/2) + \lambda_m(1 + K_m^2/2) + D}{\lambda_u\ (1 + K_u^2/2)} \right) \tag{10}$$

To modulate the polarization between the left circular and the right circular one, $\emptyset$ needs to modulate between values

$$\emptyset_1 = 2\pi(n + 1/4)\ \text{and}\ \emptyset_2 = 2\pi(m - 1/4). \tag{11.a}$$

Here n and m are arbitrary integers. For a modulation between two linear polarizations, on the other hand, the limiting values are

$$\emptyset_1 = 2\pi(n + 1/2)\ \text{and}\ \emptyset_2 = 2\pi m. \tag{11.b}$$

From Eq (8), one finds that the corresponding modulation in K_m is, in the circular case (11.a), between the values

$$K_{m1} = \sqrt{\bar{K}^2 + \Delta^2} \quad , \quad K_{m2} = \sqrt{\bar{K}^2 - \Delta^2} \quad , \tag{12}$$

where

$$\bar{K}^2 = \frac{1}{\lambda_m} (\lambda_u(1 + K_0^2/2)(p + 2m) - 2(\lambda_c + \lambda_m + D) - K_c^2 \lambda_c),$$

$$\Delta^2 = \frac{\lambda_u}{\lambda m} (1 + K_0^2/2)(p + 1/2). \tag{13}$$

Here p is another arbitrary integer. The values for the linear case (11.b) are obtained by replacing m by m + 1/4 in Eq (13).

It remains to determine the value of the corrector field K_c by requiring that the radiation intensity remains invariant as K_m varies between the values (12). When the undulator field is sinusoidal as discussed in the above, the radiation intensity becomes

$$\frac{dN}{d\Omega} = \alpha \frac{\Delta \lambda}{\lambda} \frac{4\gamma^2 K_0^2 N^2}{(1 + K_0^2/2)} G(K_c, K_m) J^2(1, q_u), \tag{14}$$

where

$$G(K_c, K_m) = \left| 1 + r_c e^{i\Theta_c} + r_m e^{i\Theta_m} \right|^2, \tag{15}$$

$$r_i = \frac{1}{N} \frac{K_i \lambda_i}{K_u \lambda_u} \frac{J(v_i, q_i)}{J(1, q_u)}, \quad (i = c, m)$$

$$J(v, q) = \frac{1}{\pi} \int_0^{\pi} \sin \phi \, \sin(v\phi - q \sin 2\phi) d\phi,$$

$$q_i = \frac{K_i^2 v_i}{4(1 + K_i^2/2)} \quad (i = u, c, m)$$

$$v_i = \frac{1 + K_i^2/2}{1 + K_0^2/2} \quad (i = u, c, m)$$

$$\Theta_c = \pi \frac{2D_1 + \lambda_c(1 + K_c^2/2)}{\lambda_u(1 + K_0^2/2)}$$

$$\Theta_m = \pi \frac{2(D_1 + D_2) + \lambda_m(1 + K_m^2/2) + 2\lambda_c(1 + K_c^2/2)}{\lambda_u(1 + K_0^2/2)}$$

The equation that determines K_c is, therefore,

$$G(K_c, K_{m1}) - G(K_c, K_{m2}) = 0 \tag{16}$$

Eq (16) can be solved numerically.

The design of a crossed undulator system proceeds as follows:
First, the optimum values of K_u and λ_u for a given magnet gap
are selected to cover the desired spectral range. Then the set of
equations, Eq (12) and Eq (16), are solved to obtain K_{m1}, K_{m2} and
K_c. Given the magnetic period lengths λ_m and λ_c, which
could be chosen to be equal to λ_u, the peak magnetic field
strengths B_i in each magnetic part are determined. The process may
need to be iterated if the resulting B_i's are not consistent with
the gap requirement and λ_i.

The time dependence of the modulator field is given by

$$B_m(t) = B_m + \Delta B_m \, f(t) \tag{17}$$

where $B_m = (B_{m1} + B_{m2})/2$, $\Delta B_m = (B_{m1} - B_{m2})/2$ and $f(t)$
is a periodic function which varies between -1 and $+1$. Ideally, $f(t)$
is a sequence of step functions. In practice it will vary smoothly
between -1 and 1, and it may be necessary to discriminate experimental
data taken when $f(t) \neq \pm 1$ by a suitable method.

IV Depolarizing Effects

In section (V), electrons are assumed to travel on an ideal
orbit. In general, the finite angular spread of electron beam as well
as the finite monochromator bandpass implies that the observed
radiation is not in a pure polarization state. To study this effect,
consider an electron which enters the undulator with a slope â with
respect to the z-axis, with an energy deviation $\delta\gamma$ relative to the
ideal energy. Let the wave length of the observed photon be $\lambda_1 +$
$\delta\lambda$, where $\delta\lambda$ lies within the monochromator bandpass. The
polarization phase ø in this case is

$$\text{ø} = <\text{ø}> - \text{ø}_o \frac{\Delta\lambda}{\lambda} + \frac{\pi L}{\lambda} (\theta^2 - <\theta^2>). \tag{18}$$

Here the angular bracket $< >$ indicates the average value, ø_o is the
phase of an ideal electron given by Eq (10), $L = 2(N\lambda_u + \lambda_c +$
$\lambda_m + D)$ is the full length of the crossed undulator system, and

$$\frac{\Delta\lambda}{\lambda} = \frac{\delta\lambda}{\lambda} + \frac{\delta\gamma}{\gamma} . \tag{19}$$

Eq (18) implies that ø has a probability distribution around its
average. The polarization property in this case can be studied using
the formalism of the coherency matrix[6], which is defined as
follows:

$$J = \begin{pmatrix} <E_x E_x^*> & <E_x E_y^*> \\ <E_y E_x^*> & <E_y E_y^*> \end{pmatrix}$$

$$
= \ |E_0|^2 \begin{pmatrix} 1 & \langle e^{-i\phi} \rangle \\ \langle e^{i\phi} \rangle & 1 \end{pmatrix} . \tag{20}
$$

Thus it is only necessary to evaluate the average value $e^{i\phi}$ to completely characterize the polarization state. If one defines the complex number u by

$$
u = \langle e^{-i(\phi - \langle \phi \rangle)} \rangle = u\, e^{i\alpha} , \tag{21}
$$

one easily obtains

$$
J = |E_0^2| \left[(1 - |u|) \begin{pmatrix} 1 & 0 \\ 0 & 1 \end{pmatrix} 1 + |u| \begin{pmatrix} 1 & e^{-i\psi} \\ e^{i\psi} & 1 \end{pmatrix} \right] \tag{22}
$$

Where $\psi = \alpha + \langle \phi \rangle$ Eq (22) is a decomposition of the radiation into its unpolarized part and completely polarized part u . The latter part has the polarization vector

$$
\underset{\sim}{e} = \frac{1}{2}(\hat{x} + e^{i\psi}\hat{y}) \tag{23}
$$

Let us assume that the distributions in $\Delta\lambda$ and Θ are Gaussian with rms values σ_λ and σ_Θ, respectively. The averaging operation in Eq (21) is then easy to perform and one obtains

$$
u = \frac{e^{-1/2\left(\frac{\sigma_\lambda \phi}{\lambda}\right)^2}\, e^{\, i\, \frac{\pi L}{\lambda}\sigma_\Theta^2}}{\left(1 + i\, \frac{\pi L}{\lambda}\sigma_\Theta^2\right)^{1/2}} \tag{24}
$$

As noted in the above, this number completely specifies the polarization state. In particular, the degree of polarization is (7)

$$
P = |u| = \frac{e^{-1/2\left(\frac{\sigma_\lambda \phi_0}{\lambda}\right)^2}}{(1 + (\frac{\pi L}{\lambda})^2 \sigma_\Theta^4)^{1/4}} . \tag{25}
$$

In order to obtain a significant polarization, the following conditions must be satisfied:

$$
\frac{\sigma_\lambda}{\lambda} = \left((\frac{\delta\lambda}{\lambda})^2 + (\frac{2\delta\gamma}{\gamma})^2\right)^{1/2} \ll \frac{1}{\phi_0} \sim \frac{1}{2\pi N} , \tag{26}
$$

$$
\sigma_\Theta > \sqrt{\frac{\lambda}{L}} \tag{27}
$$

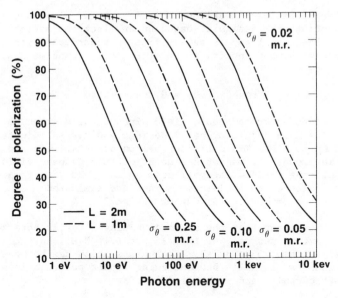

Fig. (4) Polarization From Crossed Undulator

Neglecting the electron beam energy spread, the condition (26) says
that the relative monochromator bandpass must be smaller than N^{-1},
in agreement with the discussion in Sec. (II).

     In general the inequality (26) is easy to satisfy. More serious
is the effect of electron beam angular divergence σ_Θ. Fig.
(4) shows the degree of polarization as a function of photon energy
for different values of σ_Θ, assuming $\sigma_\lambda/\lambda = 0$. Two sets of
graphs are shown, one for L = 2 meters (solid lines) the other for L =
1 meter (dashed lines). The necessity of a low emittance electron
storage ring to operate a crossed undulator system, especially at
photon energies higher than 100 eV, is clear from this figure.

(V) An Example Design for VUV Ring at NSLS

     The VUV ring [8] at NSLS runs at an electron energy of 750 MeV.
The beam angular divergence σ_Θ is 0.1 m.r. at straight section.
Thus one sees from Fig. (4) that the degree of polarization lies in
the range between 80 to 90% for photon energies between 10 eV and 30
eV for a 2 m device. Thus a crossed undulator system placed at the
VUV ring at NSLS would be a good source of VUV radiation with
arbitrarily adjustable polarization.

In order to implement various control features discussed in the previous section, the design is based on an electromagnetic structure. The magnet gap is set at 7 cm in accordance with the current vacuum chamber specification of the ring. The magnet dimensions are

$$\lambda_u = \lambda_c = \lambda_m = 15 \text{ cm}, \ D_1 = D_2 = 0, \ D_3 = 10 \text{ cm}.$$

These values ensure adequate field strength at desired photon energies. The main undulator has 3 periods and has an adjustable dc peak field up to 3 KG. The corrector has an adjustable dc peak field of 3 KG. The modulator magnet has an adjustable dc peak field of 3 KG plus a modulating field with amplitudes up to .5 KG at a frequency of 50 H_z. The poles of the modulator magnet are laminated. Fig. (5) gives a layout of the magnet design[9].

Calculation shows that the crossed undulator design here generates about 10^{14} photons per sec per $(\text{m.r.})^2$ per (.1% B.W.) in the photon energy range 10 to 30 eV. The polarization can be modulated either between two mutually perpendicular linear polarizations, or between left circular and right circular.

The design example considered here was constrained by the 7 cm vacuum chamber gap. In the future, the gap dimension could be reduced to 4 cm, and a more compact and powerful device covering a wider range of photon energies could be designed. A similar design should also work for the Aladdin storage ring at Wisconsin[10].

(VI) Discussions

The crossed undulator system studied here is a versatile source of polarized radiation for photon energies higher than 10 eV. The device works only in conjunction with a low emittance electron storage ring.

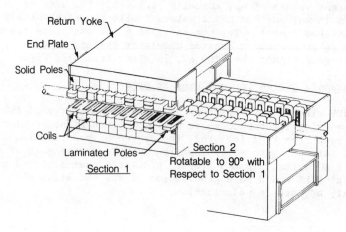

Fig. (5) A Crossed Undulator Design for VUV Ring

In the VUV energy range, (photon energies below 100eV), the current generation of VUV rings, such as Aladdin and the 750 MeV ring at NSLS, can serve as sites for the crossed undulator device. For photons of higher energy, a next generation storage ring such as the ALS[11] is required.

In general, the shape of the polarization ellipse will change due to reflections from optical elements, such as mirrors and gratings. In the case of mirrors, the effect should be small both for the low energy photons ($\epsilon \leq 30$eV) with normal incidence optics, and also for the high energy cases ($\epsilon \geq 200$eV) for sufficiently grazing angles of incidence (θ_g). In any case, it should be possible to correct for the anticipated change of the polarization due to reflections, since the source is capable of producing an arbitrary state of polarization. The case of gratings is more complicated. It is known that if

$$\frac{\lambda}{d \sin \theta_g} < .2 \tag{28}$$

then gratings behave similarly to mirrors as far as the polarization effects are concerned[12]. In the above, d is the groove spacing and θ_g is the angle of incidence. The inequality (28) is usually satisfied for VUV and soft x-ray grating systems.

The crossed undulator system works as a free electron laser if placed in an optical cavity. The operational principle here is similar to the case of the optical klystron[13], and is discussed elsewhere.[14]

Acknowledgments

I thank M. Howells for a discussion about polarization effects of gratings. This work was supported by the Office of Basic Energy Science, U.S. Department of Energy, under Contract Number DE-AC03-76SF00098.

References and Footnotes

(1) H. Winick and S. Doniach, Synchrotron Radiation Research, (Plenum Press, New York, 1980).

(2) K-J. Kim, "A Synchrotron Radiation Source with Arbitrarily Adjustable Elliptical Polarization", New Rings Workshop, SSRL Report (Stanford, August 1983); also Nucl. Instr. Meth 219, 425 (1984).

(3) R.P. Feynman, R.B. Leighton and M. Sands, The Feynman Lectures on Physics, pages 34-5, (Addison-Wesley, 1963).

(4) J.D. Jackson, Classical Electrodynamics, (John Wiley and Sons, Inc., (1962).

(5) See, for example, S. Krinsky, IEEE, NS-30 3078 (1983).

(6) M. Born and E. Wolf, Principles of Optics, (Perganon Press, 1980).

(7) Eq(25)generalizes the result of ref (2), where only terms up to
     second order in $\alpha\lambda/\lambda$ and $\alpha\theta$ are retained.

(8) NSLS publication, _Parameters_, (1983).

(9) The magnet design here is due to E. Hoyer.

(10) E.M. Rowe, et al, IEEE, NS 28, 3145 (1981).

(11) Advanced Light Source Conceptual Design Report, LBL Pub. 5084,
     Lawrence Berkeley Laboratory (1983); see also, R.C. Sah, IEEE,
     NS - 30, (1983)

(12) E.G. Loewen, N. Neviere and D. Maystre, Applied Optics 16, 2711
     (1977).

(13) N.A. Vinokurov and A.N. Skrinsky, Preprint INP 77-59, Novosibirsk
     (1977); N.A. Vinokurov, Proc. 10th Int. Conf. on High Energy
     Charged Particle Accelerators, Serpukhov, Vol. 2, 454 (1977).

(14) K-J. Kim, in _Free Electron Generation of Extreme Ultraviolet
     Coherent Radiation_, J.M.J Madey and C. Pellegrini, ed., AIP
     Conference Proceedings No. 118, page 229 (1983).

STIMULATED SYNCHROTRON RADIATION:

THE FREE-ELECTRON LASER

J. Gea-Banacloche and Marlan O. Scully
Center for Advanced Studies
University of New Mexico
Albuquerque, New Mexico 87131
and
Max-Planck Institut fur Quantenoptik
D-8046 Garching bei Munchen
West Germany

The free-electron laser is one of the most recent sources of coherent radiation. Its operation was first demonstrated at Stanford, by Madey and co-workers [1], using an electron beam from the Stanford linear superconducting accelerator; the laser wavelength in this experiment was 3.4 μm. Since then, free-electron lasers have been operated at Orsay (using a storage ring, at a laser wavelength tunable between 6350 and 6600 Å) [2], Los Alamos (using a linear accelerator, at wavelengths between 9 and 11 μm) [3a], and the University of California at Santa Barbara (using a Van der Graaf electrostatic accelerator, at a wavelength of 380 μm) [3b]. (A list of proposed experiments may be found in [4]).

The main assets of the free-electron laser are its tunability, and possible high efficiency and peak power (a peak output power of 5 MW, and average power of 3 kW has been obtained in the Los Alamos laser). It is generally agreed that this would make it a very useful source of coherent radiation in the far infrared; from the medium infrared to the near UV, a number of very good laser sources exist, and it would be a challenge for an FEL to out-perform them.

On the other hand, attempts are now being made to extend the region of FEL operation to very short wavelengths, into the vacuum ultraviolet and maybe the soft x-ray region. A number of schemes have been proposed for this [5]; those currently being under construction would use a high-energy electron beam from a storage ring [6].

33

In this paper, we will present a short review of basic FEL theory (first section), and then we will discuss some of the problems arising when one tries to go to short wavelengths, as well as some of the methods that could be used for this (second section).

The Free-Electron Laser Concept

In order to get the electrons to radiate in a free-electron laser, they are sent through a periodic, static magnetic field produced by an "undulator" or "wiggler". It is well known [7] that relativistic electrons traversing an undulator may generate very intense, tunable radiation. The radiation spectrum is peaked at a wavelength

$$\lambda_s = \frac{\lambda_w}{2\gamma_0^2} (1 + K^2) \tag{1}$$

where $\gamma_0 mc^2$ is the energy of the electrons, λ_w is the spatial period of the wiggler, and

$$K = \frac{e \, B_{rms} \, \lambda_w}{2\pi mc} \tag{2}$$

is called the "wiggler parameter" or "undulator parameter"; B_{rms} is the root-mean square value of the magnetic field of the wiggler. Equation (1) gives the fundamental frequency on axis; harmonics are also generated, usually, and the spectrum also shifts when one looks at radiation emitted off-axis.

The emission of radiation at the wavelength (1) may be understood by looking at the classical trajectories of the electrons, which are caused to "wiggle" around the direction of their initial rectilinear motion, by the periodic magnetic field. One alternative point of view, that proved useful in the development of FEL theory, is that of the Weizsacker-Williams approximation. This is based on the observation that, to an electron traveling at a relativistic speed, the static magnetic field of the wiggler appears to be very similar to a plane electromagnetic wave, of frequency $\omega' = 2\pi c\gamma_0/\lambda_w$, coming towards it (this is easily seen by performing a Lorentz translation to the electron's rest frame). One may describe then the process as Thompson scattering of this wave by the electron, in its rest frame; when a new Lorentz transformation is performed, it is easy to see that the radiation backscattered in the laboratory frame has a wavelength given by Eq. (1).

As mentioned before, the radiation emitted in this way can be very intense, more, in general, than the ordinary synchrotron

radiation emitted by conventional devices. It is, however - like
ordinary synchrotron radiation - incoherent: indeed, each electron emits completely independently of the others, with a random
phase that is related to the time of its entrance in the wiggler.
The contributions of all the electrons, therefore, add up incoh-
erently.

The free-electron laser is based on the realization by J. M.
J. Madey [8] that it is possible to obtain stimulated, coherent
emission from a device just like the one we have been discussing.
Madey's original calculation was done in an entirely quantum-
mechanical formalism, although it led to an expression for the
small-signal gain that was entirely classical, i.e., independent
of Planck's constant $\hbar$. It was not until some time later that a
classical theory of the free-electron laser was presented [9],
which led also to an understanding of the gain (and also the
saturation) mechanism.

To this end, one has to consider the motion of the electron
in the combined fields of the wiggler and of the electromagnetic
wave one wants to amplify (see Fig. 1). (Once the existence of a
mechanism for amplification has been established, a laser may be
built by enclosing the interaction region in an optical cavity and
letting the spontaneous radiation be amplified in successive round
trips, until coherence develops). At this point, it is useful to
remember what was said before, that in its rest frame the electron
sees the wiggler field as very approximately equal to that of a c
counterpropagating electromagnetic wave, of frequency $\omega' = 2\pi c \gamma_0 / \lambda_w$;
this would correspond to a wave of frequency $\omega_i \simeq \pi c / \lambda_w$ in the
laboratory frame (i.e., a wavelength $\lambda_i = 2\lambda_w$). The static wiggler
may then, to a good approximation, be replaced by such a wave.
The interference between this "wiggler wave" (or "pump wave"),
traveling in the direction opposite the electron's motion, and the
laser wave (the coherent wave one wants to amplify) results in a
sinusoidal potential that moves at close to the speed of light,
called the "ponderomotive potential" (see Fig. 2).

To be precise, if the laser frequency and wave vector are ω
and k, the ponderomotive potential is proportional to

$$\cos((\omega - \omega_i)t - (k + k_i)z) \tag{3}$$

where $\omega_i = ck_i = \pi c / \lambda_w$, and the electrons are taken to be traveling along the positive z direction. The velocity of this
"ponderomotive wave" is clearly

$$v_s = \frac{\omega - \omega_i}{k + k_i} = c\,\frac{1 - k_i/k}{1 + k_i/k} \tag{4}$$

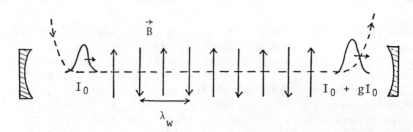

Fig. 1.   Scheme of a free-electron laser. The dashed line repre-
          sents the trajectory of the electron beam. The pulse of
          coherent radiation of intensity I_0 is amplified by gI_0
          (where g is the gain factor) after interacting with the
          electrons in the wiggler.

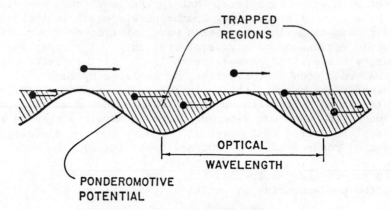

Fig. 2.   The ponderomotive potential. Electrons are trapped in
          its wells, or not, depending on their initial phases. In
          any case, their motion is given by simple pendulum
          equations.

This is indeed very close to c for ultrarelativistic electrons, since k_i/k is of the order of $1/\gamma^2$ (compare Eq. (1); one expects $k \sim k_s$, that is, the laser wavelength should be close to the peak of the spontaneous emission spectrum).

The electrons perform longitudinal oscillations in this ponderomotive potential; the equations describing their motion are like those for a simple pendulum. Initially, their phases are all random but, after some time, they tend to accumulate in the potential wells. In this way the electron beam, initially uniform, becomes bunched at the optical wavelength; a phase relationship is thus established between the electrons, and they radiate now coherently, amplifying the injected signal.

To be more precise, it is found that, if the initial velocity of the electrons is slightly larger than the velocity of the ponderomotive wave (v_s), they indeed amplify the injected signal, but if their velocity is smaller than v_s they absorb energy from the wave, and no gain (rather, absorption) results. This may be understood intuitively in terms of those electrons moving faster than the potential wave giving up energy, and conversely for those going more slowly. This situation is familiar in the context of Landau damping, and it is sometimes stated that the gain mechanism in a free-electron laser is a kind of inverse Landau damping. Note that, by conservation of energy, the energy lost by the electrons has to go into amplification of the injected wave, and conversely, if the electrons gain energy it has to be by depleting the injected wave (the wiggler or pump wave has constant energy, since it is really a static field).

It is not hard to see that the velocity of the electrons exactly coincides with that of the ponderomotive wave, only if the wavelength of the injected signal equals the λ_s of Eq. (1). This means that at exactly the peak of the spontaneous emission spectrum there is neither gain nor absorption. For a given electron energy, gain would take place at wavelengths slightly larger than λ_s, while gain at a given wavelength takes place for a slightly larger electron energy than that giving peak spontaneous emission at that wavelength. This is summarized in Fig. 3, which shows the well-known antisymmetric small-signal gain curve for the free-electron laser. It is interesting to note that this curve is modified when the gain per pass is large (larger than 100%); in particular, it no longer is antisymmetric (see, for instance, [10]). It would also be modified if quantum effects became important: this is the case when the energy of the photons emitted is larger than the width of the curve in Fig. 3 (expressed in units of electron energy instead of velocity); this possibility is treated in [11].

The maximum achievable gain in the classical, small-gain-per pass limit is given by the formula

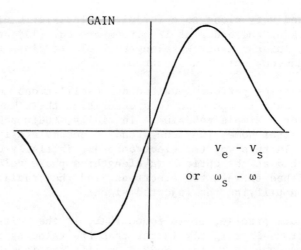

Fig. 3. The gain in the free-electron laser in the small-signal,
         small-gain-per-pass regime, plotted as a function of
         either the velocity detuning (for gain at a given wave-
         length) or the frequency detuning (from the frequency
         of spontaneous emission for a given electron energy).

$$g_{max} = 3.05 \; \frac{\lambda^{3/2} \lambda_w^{1/2}}{\Sigma} \; \frac{K^2}{(1 + K^2)^{3/2}} \; N^3 \; \frac{I}{I_A} \tag{5}$$

where Σ is the average laser beam cross section, N the number of
wiggler periods, I is the (peak) current in the electron beam, and
I_A = $ec/r_0 \simeq$ 17,000 amps (r_0 is the classical electron radius).
Lest it seem that the gain may be increased arbitrarily by decreasing the cross-section of the laser beam, it has to be pointed out
that nothing can be gained by making it smaller than the cross-
section of the electron beam; when both are approximately equal
one has optimum coupling, and in that case it may be seen that the
gain is really proportional to the current density in the electron
beam. There are important factors limiting the current densities
that can be achieved; notably space charge effects and, most
importantly, the finite emittance of the beam. This is to be kept
in mind when considering the possible operation of the free-
electron laser at very short wavelengths.

Fels in the VUV and Soft X-Ray Region

     As was mentioned in the introduction, there has been recently

a surge of interest in short wavelength FELs. The difficulties
associated with going into that region of operation are clear from
the expression (5) for the small-signal gain, since the gain
scales as the laser wavelength λ to the 3/2; therefore, going from
the IR to the UV the gain may be reduced by one or two orders of
magnitude. This is particularly serious because of the lack of
good mirrors at short wavelengths: in order to obtain lasing
action the gain has to exceed the losses at the mirrors. This
means that gains per pass as high as unity (100% or more) are
necessary.

According to Eq. (5) there are only two practical ways to
achieve this: increasing the current density in the electron beam
and increasing the number of wiggler periods. In a conventional
wiggler, the period λ_w cannot in practice be made much smaller
than, say, 1 cm. Hence, the need for a larger number of wiggler
periods leads to the consideration of very long wigglers (around
20 meters long; see [5] and [6]). These present several problems:
alignment problems, high cost (if they are made with permanent
magnets, the number of magnets needed is, of course, proportional
to the number of wiggler periods) and the requirement of a very
small beam emittance, in order to keep the electron beam focused
to a small radius over the increased interaction length.

These and other difficulties were analyzed in several of the
papers in the book quoted in Ref. 5, and it appears that, despite
all, FEL operation at wavelengths as short as 100 Å might be
achievable with long wigglers and storage rings. The latter have
to be specially built, since a long straight section is necessary
to accomodate the wiggler; some of these projects are already
underway [6].

An alternative to the use of very long conventional wigglers
(suggested in [12]) is that of replacing the wiggler by an actual
electromagnetic wave, of very high intensity, such as may be
obtained from a high-power laser (Nd:glass, for example). That
such a replacement is possible in principle is clear from the fact
emphasized above that the static magnetic field of the wiggler
looks to the electrons like the field of a counterpropagating
electromagnetic wave of wavelength $2\lambda_w$, and in fact, the latter
may be substituted for the former, to a very good approximation,
in the theoretical calculations. The advantage of this so-called
"optical" (or electromagnetic-wave) wiggler is that its period may
be made very small (the wavelength of the radiation from a
Nd:glass laser is 1.06 μm), so that a very large number of periods
may be accomodated in a comparatively small interaction region.
It also means that one doesn't have to buy a few thousand permanent magnets, although the cost of the very-high-power lasers that
could supply an "optical wiggler" pulse is also rather high. The

possibility of using optical wigglers to obtain coherent radiation
at wavelengths much shorter than any now envisioned (around 6 A)
has been studied, from different perspectives, in [12] and [13].
In [12], a non-colinear geometry was proposed, (see Fig. 4), and
the 6 Å radiation would be enclosed in a ring resonator using
Bragg reflection off mirrors of monocrystalline germanium. Gain
and mirror reflectivity calculations are presented there, and it
is shown that a small-signal gain equal to 6 (600%) per roundtrip
could be achieved (this is large enough to make it necessary to
perform a more exact calculation for the so-called "high-gain
regime" [10,13]; the actual factor by which the laser radiation
would be amplified in each pass would be somewhat larger than 6);
the total losses per roundtrip could probably be kept smaller than
10%, in a narrow range of wavelengths, so that in principle it
should be possible to bring the system above threshold.

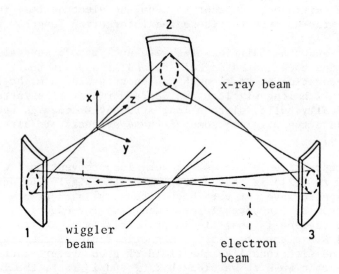

Fig. 4.   The scheme for an x-ray FEL proposed in Ref. 12. The
          static wiggler is replaced by a high-intensity radiation
          pulse (labeled "wiggler beam" in the figure) that meets
          the electron beam at an angle (greatly exaggerated in the
          figure, for clarity). The resonator uses Bragg
          reflection to contain the x-rays in a ring cavity.

One has to note, however, that both in [12] and [11], the
electron beam is supposed to satisfy some very stringent requirements, especially as regards peak current, energy spread, and
emittance. The values assumed for the last two, in particular,
are at the edge of what can be achieved with current technology;
of the accelerators being used today in free-electron lasers, only
the electrostatic accelerator at UCSB [3] comes close to meeting
them, and its peak current is unfortunately very low. However,
accelerators are being built today specifically for free-electron
laser applications, and improvements in design and technology
might make free-electron generation of coherent soft x-ray radiation possible in the not-too-distant future.

REFERENCES

1. D. A. G. Deacon, L. R. Elias, J. M. J. Madey, G. J. Ramian,
    H. A. Schwettman, and T. I. Smith, Phys. Rev. Lett. 38,
    (1977).
2. M. Billardon, P. Elleaume, J. M. Ortega, C. Brazin, M.
    Bergher, Y. Petroff, M. Belghe, D. A. G. Deacon, J. M. J.
    Madey, in Free-Electron Generators of Coherent Radiation,
    C. A. Brau, S. F. Jacobs, M. O. Scully, eds., Proc. SPIE
    453 (1984) p. 269.
3. a) B. E. Newman, R. W. Warren, J. C. Goldstein, and C. A.
    Brau; b) L. R. Elias, in Proceedings of the 1984
    Free-Electron Laser Conference, to be published in J.
    Nucl. Inst. Meth.
4. P. Sprangle and T. Coffey, Physics Today, Vol. 37, No. 3, p.
    44 (1984).
5. See the book Free-Electron Generation of Extreme Ultraviolet
    Coherent Radiation, J. M. J. Madey and C. Pellegrini,
    eds., AIP Conference Proceedings No. 118, (1984).
6. See the paper by J. M. Peterson et al. in the Proceedings
    quoted in Ref. [3].
7. A. Bienenstock and H. Winick, Physics Today, Vol. 36, No. 6,
    p. 48 (1983).
8. J. M. J. Madey, J. Appl. Phys. 42, 1906 (1971).
9. F. A. Hopf, P. Meystre, M. O. Scully, and W. H. Louisell,
    Opt. Commun. 18, 413 (1976).
10. C. Shih, in the book quoted in Ref. [2], p. 205.
11: J. Gea-Banacloche, to be published in Phys. Rev. A.
12. P. Dobiasch, P. Meystre, and M. O. Scully, IEEE J. Quantum
    Elec. QE-19, 1812 (1983).
13. J. Gea-Banacloche, G. T. Moore, and M. O. Scully, in the book
    quoted in Ref. [2], p. 393.

NATURAL AND MAGNETIC VACUUM ULTRAVIOLET CIRCULAR DICHROISM MEASUREMENTS

AT THE SYNCHROTRON RADIATION CENTER UNIVERSITY OF WISCONSIN-MADISON

Patricia Ann Snyder, Paul N. Schatz, and Ednor M. Rowe

Dept. of Chemistry, Florida Atlantic Univ., Boca Raton, FL 33431
Dept. of Chemistry, Univ. of Virginia, Charlottesville, VA 22901
Synchrotron Radiation CTR., Univ. of Wisconsin-Madison
Stoughton, Wis. 53589

INTRODUCTION

Substances which rotate the plane of polarization of linearly polarized light are called optically active. The optical rotation occurs due to unequal responses to left and right circularly polarized radiation. Circular dichroism spectroscopy measures the difference in absorption between left and right circularly polarized light (usually expressed as the difference in molar extinction coefficient, $\varepsilon_L - \varepsilon_R$, or absorbance, $A_L - A_R$) as a function of wavelength.

A molecule with no plane or center of symmetry will exhibit optical rotation and a circular dichroism spectrum in the absence of external influences. This is called natural circular dichroism. Molecules which normally do not rotate linearly polarized light will do so in a magnetic field and have a magnetic circular dichroism spectrum.

The signal to noise ratio for natural and magnetic circular dichroism is statistical and proportional to the number of photons detected by the photomultiplier.[1] As a result, circular dichroism measurements have been difficult or impossible in certain regions of the spectrum. The development of a calcium fluoride stress plate modulator,[2-4] the magnesium fluoride Wollaston polarizer[5] and modifications allowing more intensity from Hinteregger lamps [6,7] all helped make circular dichroism measurements possible in the vacuum ultraviolet region.[7-18] Vacuum ultraviolet circular dichroism measurements with a conventional light source have been difficult, time consuming, and limited in both their resolution and energy capabilities. However, these measurements have clearly demonstrated the usefullness of measurements in the vacuum ultraviolet region. (For example, see references 7 to 32.)

Snychrotron radiation from modern electron storage rings is highly linearly polarized, a continuum, collimated and more intense than conventional vacuum ultraviolet sources. These attributes are ideal

for circular dichroism measurements. Circular dichroism instruments
which use synchrotron radiation have been developed and are already
contributing to our knowledge. The location, design, and status of the
various natural and magnetic circular dichroism instruments, which use
synchrotron radiation, has been summarized in a recent paper.[33]   In
this paper, the present instrumentation for magnetic and natural
circular dichroism measurements at the Synchrotron Radiation Center,
University of Wisconsin-Madison, will be discussed first. Next the
natural and magnetic circular dichroism results which have been
obtained with this instrument will be summarized, and then the future
plans for instrumentation at the Synchrotron Radiation Center,
University of Wisconsin-Madison will be discussed briefly.

INSTRUMENTATION

     In order to preserve and enhance the linear polarization of the
synchrotron radiation, all reflections are s reflections. As a result,
a degree of linear polarization of 99% is expected at the entrance to
the experimental chamber with the monochromator slits wide open.
Therefore, no polarizer is necessary with the present beam line,
monochromator, and experimental chamber design. This has two
advantages. First, there is more light intensity without a polarizer.
More intensity means a better signal to noise ratio and therefore
better resolution as well as faster scans are possible. The better
resolution is particularly important for magnetic circular dichroism
because it is necessary for proper interpretation of the data. Second,
measurements can be made to higher energies, since it is the magnesium
fluoride polarizer which has limited measurements to 135 nm. Both
natural and magnetic circular dichroism measurements have been made to
125 nm at the Synchrotron Radiation Center, University of
Wisconsin-Madison. The present wavelength limit is due to the calcium
fluoride quarter wave modulator.

     The following is a short description of the optics of the
system.[34] (Fig. 1 shows a schematic of the natural circular
dichroism instrument starting with the mirror at the entrance of the
monochromator.[33,35-36])

     Twenty-five milliradians of horizontal beam and five milliradians
of beam above and below the plane of the electron orbit are collected
by a 5° grazing angle ellipsoidal mirror which produces a 1:1 image of
the synchrotron radiation source. This upward reflection is followed
by a 5° grazing angle downward reflection from a plane mirror resulting
in the light beam leaving the two mirror system parrallel to the floor.
The light then impinges on a gold coated cylindrical mirror placed
before the entrance slit of the one meter Seya-Namioka monochromator.
The light then passes through the entrance slit to the 1200 ℓ/mm magnesium fluoride coated grating (8.3A/mm dispersion) and out the exit
slit to a gold coated cylindrical mirror which focuses the light
beam in the experimental chamber. Since the angles of convergence to
the focus are small, the beam maintains its small size and can be put
through the small bore of the super conducting solenoid without
additional optics. This is an important consideration, since the
addition of each optical element decreases the intensity. The optical
system has been designed to maximize the intensity and as a result it

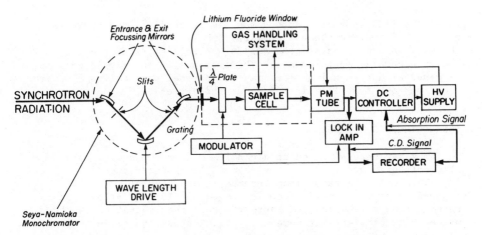

Fig. 1. A schematic of the instrument.

is possible to make magnetic circular dichroism measurements with a
.04nm spectral bandwidth. This bandwidth limit is due to the
Seya-Namioka monochromator, not a lack of light intensity.

When the beam leaves the monochromator it passes through a lithium
fluoride window which separates the beam line and monochromator from
the experimental vacuum chamber. The light then passes through a split
head calcium fluoride quarter wave modulator (Hinds Inc) which has its
fast axis at 45° to the plane of the linear polarized light. The
quarter wave modulator operates at 50k Hz, alternately making left and
right circularly polarized light. The modulated light then passes thru
the sample cell and proceeds to the photomultiplier tube (EMI 9635B).
Since the photomultiplier is pyrex, it is coated with sodium salicylate
which fluoresces at 400nm. With an optically active sample, the
photomultiplier sees a DC signal upon which an AC signal is
superimposed. To a good approximation the circular dichroism is
proportional to the AC signal divided by the DC signal[1]

$$A_L(\lambda) - A_R(\lambda) = \text{Constant} \frac{AC}{DC}$$

The DC signal is kept constant by the DC controller which varies the
high voltage to the photomultiplier. The AC signal is then
proportional to the circular dichroism. The magnitude and sign of the
AC signal is detected with a lock-in amplifier and then outputted to
one pen of a two pen chart recorder. The other pen follows the change
in the voltage supplied to the photomultiplier and is related to the
absorption of light by the sample. The sign and magnitude of the
proportionality constant is determined by measuring a standard, such as
d-10-camphorsulfonic acid.[37] The instrument, as described, is for
natural circular dichroism measurements. For magnetic circular
dichroism measurements a superconducting magnet with a room temperature

bore (7 T field) is placed around the sample cell.[38] Measurements
can be carried out in the gas or liquid phase for natural circular
dichroism. At the present time only a gas cell exists for magnetic
circular dichroism measurements. The pressure in the sample cell is
determined with a capacitance manometer (barocel sensor type 570 with
analog read out type 1173 from Dresser Industries).

NATURAL CIRCULAR DICHROISM

The first natural circular dichroism measurements with this
instrument were carried out on (+)-3-methylcyclopentanone.[33,
35-36] This molecule was chosen because it has been a
"standard" test molecule for new instruments. It thus allows the instrument to be tested and the performance of the instrument can be
compared with previous instruments. Fig. 2 shows the circular dichro-

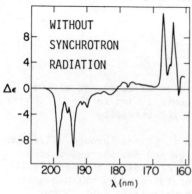

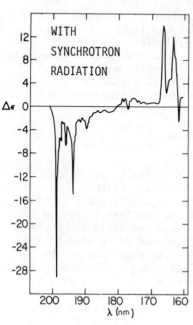

Fig. 2. Natural circular dichroism
spectrum of (+)-3-methyl-
cyclopentanone in the vapor
phase between 200-160nm
without and with synchrotron
radiation.

ism spectrum from 200nm to 160nm of gaseous (+)-3-methylcyclopentanone
obtained with synchrotron radiation compared [33,35-36]
with the best results obtained without sychrotron radiation.[7] The
spectral bandwidth with synchrotron radiation was .17A, and the time
constant 4s. Therefore, not only is the data better resolved with
synchrotron radiation, but it was obtained more quickly. (The shorter
the time constant the faster you can scan the spectrum.) Fig. 3 shows
the natural circular dichroism spectrum of (+)-3-methylcyclopentanone
obtained with synchrotron radiation between 170nm and 125nm. The
spectral bandwidth was .17nm from 170nm to 148nm and .41nm from 148nm
to 125nm. (Previously the best spectral bandwidth was .8nm between 160
and 135nm.[7])

    Since (+)-3-methylcyclopentanone has been used as a test molecule,
a number of circular dichroism spectra exist in the
literature.[7-9,20,33,35,36] Some work has
been done on interpreting the spectrum, but it has not yet been
interpreted in detail.[20,22,39-40] This points up
the need for theoretical work in the field of natural circular
dichroism as well as the need for additional data. The data, however,
needs to be on a series of related relatively simple molecules, i.e.
molecules which can be attacked theoretically.

MAGNETIC CIRCULAR DICHROISM

General Comments

    A magnetic circular dichroism spectrum contains information on the
electronic structure, number of transitions, and the assignment of

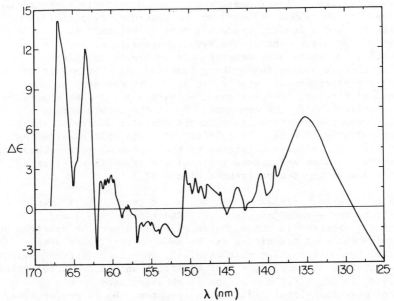

Fig. 3. Natural circular dichroism spectrum of (+)-3-methylcyclopent-
         anone between 170 and 125nm with synchrotron radiation.

transitions in a molecule. It is particularly useful when applied to molecules which have some allowed degenerate states. For magnetic circular dichroism[41,42]

$$\Delta\varepsilon = [A(\frac{\partial f}{\partial \nu}) + (B + \frac{C}{kT}) f] H$$

where f is the absorption line shape, ν the frequency, H the magnetic field and T the temperature. The three terms (A, B, and C) are each linearly dependent on the field strength, but they have different dependences on the absorption line shape and on the temperature. A brief description of each term follows.

A terms result when the ground or excited state is degenerate. An A term has a derivative line shape with a magnitude dependent on the difference in magnetic moment of the two states. A terms are temperature independent and diagnostic of a degenerate state.

B terms can be present for any transition and result from field induced mixing between states. B terms have the line shape of the absorption peak and are temperature independent.

C terms result when the ground state is paramagnetic. C terms have the line shape of the absorption peak, but they are temperature dependent.

Benzene

The initial magnetic circular dichroism measurements with the instrument at the Synchrotron Radiation Center, University of Wisconsin-Madison were carried out on gaseous benzene[38] for the following reasons. First, measurements had previously been carried out on benzene in the 180nm region[18] so the instrumental performance could be checked. Second, there has been a longstanding debate concerning the assignment of the Rydberg transitions in benzene.[42, 43] Therefore, measurements to higher energies could be used to assign the Rydberg transitions. The magnitude and sign of the experimental results for A_1/D_0 are compared to the theoretical results. For a transition with a B term the experimental and theoretical B_0/D_0 are compared. B_0 is the zeroth moment obtained from the experimental magnetic circular dichroism curve. Theoretically B_0 is more difficult to calculate than A_1 since B_0 involves all the electronic states of the system whereas A_1 depends only on the excited and ground state of the transition. (For a detailed discussion see references 41 and 42.)

Since benzene belongs to the D_{6h} point group, the allowed transitions are either $^1A_{2u}$ or $^1E_{1u}$. Therefore, both A and B terms are expected in the magnetic circular dichroism spectrum and the assignment of a transition can be made on that basis alone. In addition, for a transition which exhibits an A term further information can be obtained by comparison of theoretical and experimental values of A_1/D_0. A_1 is the first moment and is obtained experimentally from the magnetic circulat dichroism spectrum. D_0 is proportional to the dipole strength of a transition and therefore it is experimentally

related to the area under the absorption curve. It is the zeroth
moment of the absorption curve.

The absorption spectrum of benzene has four Rydberg series
labelled R, R', R" and R'" by Wilkinson.[44] While the Rydberg
series are known, their assignment has been uncertain. The magnetic
circular dichroism of benzene vapor was studied between $52630 cm^{-1}$
and $74074 cm^{-1}$ with attention focused on the region between
$67200 cm^{-1}$ and $67800 cm^{-1}$ which contains the 3R', 3R" and 3R'"
Rydberg transitions. Fig. 4 shows the magnetic circular dichroism and
absorption in this region.[38]

The magnetic circular dichroism associated with the R series
($65703 cm^{-1}$) is very small and shows no indication of an A term.
The 3R' origin shows a prominent A term. If the R and R' series are
associated with the $\pi(e_{1g}) \rightarrow np$ excitations then the R series must
correspond to the $^1A_{1g} \rightarrow {}^1A_{2u}(\pi(e_{1g}) \rightarrow np(e_{1u}))$

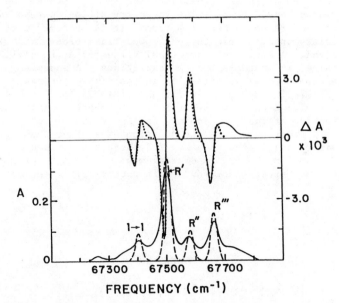

Fig. 4. Benzene vapor. The experimental absorption spectrum (lower
         solid curve) and the magnetic circular dichroism spectrum per
         Tesla (upper solid curve) at room temperature and a pressure
         of .038mm for a cell length of 13.2cm with a spectral band-
         width of .04nm. The upper dotted curve is the best gaussian
         fit of the magnetic circular dichroism spectrum and the lower
         dotted curve is the corresponding abosrption fit using the
         same E° and Δ values as for the magnetic circular dichroism.
         The time constant was 4s.

transition and the latter must correspond to the $^1A_{1g} \rightarrow ^1E_{1u}$
($\pi(e_{1g}) \rightarrow np_z(a_{2u})$) transitions. These magnetic circular
dichroism results finally settle the debate concerning the assignment
of the R and R' series.

As an additional check on the transition assignment, theoretical
calculations of A_1/D_0 were carried out for the 3R' transition, and
compared to the experimental results. To obtain a theoretical value of
A_1/D_0 a simple model which uses standard basis functions is used.
The excited state is assumed to be a Rydberg p_z orbital and the
ground state is assumed to be the highest occupied π molecular
orbital. A simple Huckel π molecular orbital is assumed for the ground
state. Using this model a positive A_1/D_0 is obtained with a
magnitude of .95. The experimental value is positive, with a 1.26
magnitude. (see Table I.) The theoretical value agrees with the
experimental value in sign and approximate order of magnitude.

The next question is the assignment of the R" and R'" series. The
magnetic circular dichroism spectrum shows what appear to be
overlapping A terms in this region of the spectrum. One possibility is
that R" and R'" are associated with $\pi(e_{1g}) \rightarrow nf$ Rydberg transitions.
The $\pi(e_{1g}) \rightarrow nf$ Rydberg transitions give rise to three dipole-allowed
transitions. Two of these transitions are to $^1E_{1u}$ states and
therefore would have A terms. The third transition is to a $^1A_{2u}$
state and therefore would have a B term. It is proposed that the R"
and R'" series contain these three allowed transitions which are not
fully resolved. A positive theoretical value for A_1/D_0 of 1.0 $\pm$
.05 is obtained for both $^1A_{1g} \rightarrow ^1E_{1u}$ transitions. A gaussian
fit of the experimental data results in positive experimental values
for A_1/D_0 of .78 for 3R" and .76 for 3R'. (See Table I) Thus the
magnetic circular dichroism spectrum lends support to this assignment
but the question of the location of the $^1A_{2u}$ state is not
apparent. In order to settle the question of whether the $^1A_{1g} \rightarrow$
$^1A_{2u}$ transition associated with the f Rydberg transitions is
present and confirm the proposed assignment of the A terms for the R"
and R'" series, better resolved data is necessary. (See the section on
future plans.)

TABLE I[38]

Line	$E_0(cm^{-1})^a$	Δ $(cm^{-1})^a$	A_1/D_0 ^{a}	A_1/D_0(calc.)
3R'	67,531	16	1.26	.95
3R"	67,609	16	.78	1.00
3R'"	67,691	16	.76	1.00

a) Parameters were obtained by gaussian fit of the magnetic circular
   dichroism and absorption using the rigid shift model.[41]

Olefins

The simplest olefin, ethylene, serves as the basis for
understanding all other olefins.[43] In the ground state ethylene is
planar and it belongs to the D_{2h} point group. Since the D_{2h}
group has no degenerate states, and ethylene is not paramagnetic, the
magnetic circular dichroism consists only of B terms. B terms are the
most difficult to interpret. Also, B terms are usually less intense,
and therefore more difficult to measure. However, since ethylene is
basic to our understanding, information on the number of transitions in
an energy region, as well as their assignment is important. Therefore,
magnetic circular dichroism measurements were carried out in the
following regions:

$$74074 \text{ cm}^{-1} \text{ to } 71428 \text{ cm}^{-1}$$
$$60606 \text{ cm}^{-1} \text{ to } 57143 \text{ cm}^{-1}$$
$$\text{and around } 69444 \text{ cm}^{-1}$$

These measurements allowed the study of the 2R, 3R, 3R" and 4R'''
Rydberg series (Wilkinson notation[45]). The magnetic circular
dichroism and absorption in the 71500 cm^{-1} to 73500 cm^{-1} region
are shown in Fig. 5.[46-47] The $3R_{00}$ and $4R_{00}'''$
Rydberg transitions are indicated on Fig. 5. The magnetic circular
dichroism spectrum clearly shows that both the $3R_{00}$ and $4R_{00}'''$
Rydberg transitions are a composite of two transitions each since there
appear to be positive A terms for these transitions. A pseudo A term
could arise from accidental degeneracy of the correct symmetry.[46]

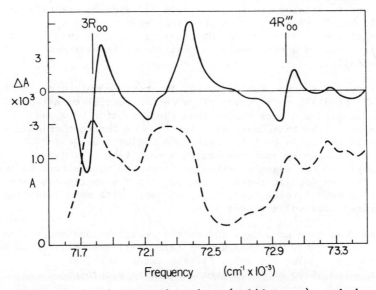

Fig. 5. The magnetic circular dichroism, (solid curve), and absorption,
(dashed curve), of ethylene in the gas phase. Spectral
bandwidth .075nm with a 12.5s time constant.

The $3R_{00}$ transition has been assigned as belonging to the s
Rydberg series, one of the dδ (B_{2u}) Rydberg series, and both of the
allowed dδ Rydberg series. (43,48-49, and references cited therein)
It seems reasonable to assume that the two allowed dδ transitions would
be 'accidentally' degenerate and since their polarizations are in the x
and y direction, a pseudo A term would be expected.[46] A pseudo A
term would not be expected for a s Rydberg transition or if only one of
the d transitions occurred in this energy region.

The $4R_{00}$"' has been assigned as s, s + ν_3, dπ_x and dσ (43,
48-49 and references cited therein). None of these assignments would
be expected to exhibit a pseudo A term. The magnetic circular dichroism shows that there is either another electronic transition in this
energy region or previously unassigned vibrational structure. [46]

Since the magnetic circular dichroism of the $4R_{00}$"' transition
has a pseudo A term of the same sign as the $3R_{00}$ transition, and is
located 1233 cm^{-1} to the blue of the $3R_{00}$ transition, we
propose that the $4R_{00}$"' transition is the first member of an a_g
vibrational progression for the dδ ($3R_{oo}$) Rydberg transitions. The
ν_3 vibration is of the correct symmetry and approximate magnitude.
(Another possibility would be ν_2.)[46] These results are
consistent with McDiarmid's conclusion that the $4R_{00}$"' cannot be an
origin because analogous bands are not found in any of the isotropic
molecules of ethylene.[48] If this is the first member of a
vibrational series then the A_1/D_0 ratio should be the same as for
the zero-zero transition, as is observed.[47]

Theoretical calculations have been carried out of for the
assignment of $3R_{00}$ to both of the allowed dδ Rydberg
transitions.[47] A positive pseudo A term results. This is
consistent with experiment. The magnitude of the theoretical result is
an order of magnitude larger than the experimental result. However,
this is not surprising since very simple 3dδ Rydberg orbitals were
used. If $4R_{00}$"' is assigned as the first member of an a_g
vibrational progression for the dδ transitions, then the sign of the
theoretical pseudo A term would be positive as is observed
experimentally.

The use of synchrotron radiation for measurements made in the
54074 cm^{-1} to 51428 cm^{-1} region allow better resolution as well
as measurements to higher energy than the very nice previous
measurements of Brithlindner and Allen.[50]. As the resolution
improved in the 54074 cm^{-1} to 51428 cm^{-1} region, the second
member of the Rydberg doublets looked more A like (accidental
degeneracy) and the magnetic circular dichroism went almost to zero
between the two doublet peaks. However, these measurements were
difficult and need to be repeated, preferably with more intensity and a
computerized data collection system.

There has been some very interesting multiphoton inonization work
on ethylene recently.[51] This work located what are believed to be
the forbidden $\pi{\to}3p_y$ and $\pi{\to}3p_x$ Rydberg transitions, at
62905 cm^{-1} and 66875 cm^{-1}. Magnetic circular dichroism

measurements were made to look for these transitions but neither they nor the missing $\pi \rightarrow 3p_z$ were found.

The deuterated ethylenes have more clearly defined structure than ethylene.[45],[48] Therefore, their magnetic circular dichroism was expected to be clearer. In addition, data on the deuterated ethylene's should help sort out the vibrational structure. Preliminary measurements have been carried out on fully deuterated ethylene (C_2D_4).

Fig. 6 shows the magnetic circular dichroism and absorption in the 71,500 cm^{-1} to 75, 5000 cm^{-1} region for fully deuterated ethylene. This region includes the $3R_{00}$ (72047 cm^{-1}) and $4R_{00}''$ (73007 cm^{-1}) Rydberg transitions. As anticipated, the magnetic circular dichroism spectrum of fully deuterated ethylene is much better resolved than for ethylene. The magnetic circular dichroism spectrum proves that there are more transitions in this region than what have been observed in the absorption spectrum.

In the future the magnetic circular dichroism of deuterated ethylene will be completed and the interpretation done in conjunction with ethylene. In addition data on the other deuterated ethylenes will be obtained and should help the interpretation. When it is completed, this investigation should add considerably to our knowledge of the electronic structure of ethylene and, therefore, olefins in general.

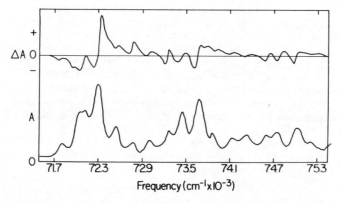

Fig. 6. The magnetic circular dichroism (upper curve) and absorption (lower curve) for fully deuterated ethylene in the vapor phase. The spectral bandwidth is .04nm, the time constant 12.5 sec and the magnetic field 6.65 Tesla.

Acetylene

The electronic spectrum of acetylene is even less well understood than that of benzene and ethylene even though it has also been the subject of numerous studies, including one of the first vacuum ultraviolet magnetic circular dichroism measurements.[52-53] However, since the use of synchrotron radiation allows measurements to higher energies as well as with better resolution, its study was undertaken. It also has the advantage of allowing a comparison of the new data with the previous data.

The data has not been reduced yet. However, the following results are clear. First, better resolution has resulted in considerable more structure and the appearance of some positive bands in the 190 - 170nm region. However, this region will still be very challenging to interpret. Second, measurements in the 155 to 139nm region are consistent with the previous work[53] in that four A terms corresponding to a $^1\pi_u$ state zero-zero transition with a progression of the C-C stretching frequency is observed. It will be interesting to compare our A_1/D_0 ratio with theory and the previous results. Third, the measurements to higher energy show clearly the presence of a second $^1\pi_u$ (one with a different A_1/D_0 ratio) state and what appears to be the first member of a C-C stretching frequency progression for this transition.

Oxygen

Oxygen belongs to the $D_{\infty h}$ point group and can have both A and B terms. Since it is paramagnetic, it also has C terms. The magnetic circular dichroism was obtained in the Schumann-Runge region $^3\Sigma_g^- \rightarrow {}^3\Sigma_u^-$ (175nm to 185nm). A small region of the spectrum is shown in Fig. 7. The spectral bandwidth is .04nm and the time constant was 1.25 sec. Preliminary calculations show that the $^3\Sigma_g^- \rightarrow {}^3\Sigma_u^-$ transition cannot give rise to A, C. or B (in state) terms even in the presence of spin-orbit coupling. Thus any magnetic circular dichroism is associated with out of state effects. The observed magnetic circular dichroism, however, is very large. Field induced $^3\pi_u$ with $^3\Sigma_u^-$ can produce B terms while spin-orbit mixing of these states produce A and C terms. Thus, the magnetic circular dichroism is a sensitive probe of the interaction of the $^3\Sigma_u^-$ and $^3\pi_u$ states. The magnetic circular dichroism increases sharply (relative to the absorption) starting with $0 \rightarrow 12$ vibronic line (55,735cm^{-1}). Thus it appears that the magnetic circular dichroism gives information on the crossing point of the $^3\pi_u$ state with the $^3\Sigma_u^-$ state.[54] These preliminary results on oxygen will be checked and the spectrum reduced. In addition, the matrix isolated magnetic circular dichroism measurements will be made (when the equipment is available) to identify the C terms.

FUTURE PLANS

In this section plans for instrumentation at the Synchrotron Radiation Center, University of Wisconsin-Madison, which will be

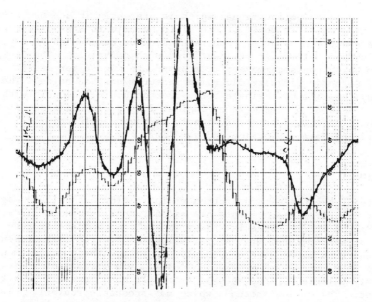

Fig. 7. The magnetic circular dichroism (darker line) and absorption
(lighter line) of oxygen as outputted to the recorder near
179nm. The spectral bandwidth was .04nm and the time constant
1.25 seconds.

carried out in the immediate future, are discussed. These
modifications have already been started or will be started within the
next year.

   All the measurements discussed in this paper were carried out with
a Seya-Namioka (one meter) monochromater. The best spectral bandwidth
obtainable when this monochromator is functioning as designed is .04nm.
From the measurements which have been carried out on benzene, it is
clear that better resolution would be useful. (Better resolution is
needed in the 3R" and 3R"' region for benzene in order to assign these
transitions.) A four meter monochromator has been designed and is
being constructed by the Synchrotron Radiation Center, University of
Wisconsin-Madison.[55] This monochromator has all s reflections, and
a possible spectral bandwidth at least ten times smaller than the
present one meter Seya-Namioka monochromator. In addition, the four
meter monochromator will have four times the intensity of the Seya-
Namioka monochromator at the same spectral bandwidth and beam current.
(This is important, since the signal to noise for circular dichroism is
proportional to the square root of the intensity.) The four meter
monochromator will be one of the first monochromators installed on
Aladdin. Aladdin is the new storage ring being built at Wisconsin.
Eventually Aladdin will run at a higher current and with more intensity
than Tantulus I. (All the measurements made in this paper were carried
out on Tantulus I.)

The superconducting magnet which is presently in use has a room temperature bore. It would be useful to be able to study the magnetic circular dichroism at other temperatures. For cases such as oxygen it will allow the C terms to be identified with certainty. The design of a system which can accommodate an appropriate magnet for matrix isolated magnetic circular dichroism measurements has begun. The magnet system will be a single bore superconducting solenoid with a possible field of 8 T at 4.2K or 10 T at 2.2 K. In addition, it is hoped that funds will become available in the near future for computerizing data collection and handling.

CONCLUSION

Already the use of synchrotron radiation for natural and magnetic circular dichroism measurements has added to our knowledge. [33, 35-36,38,46-47,56-60] Clearly, there is still much to be learned in the regions of the spectrum presently available. However, measurements to higher energies would also be useful. To extend these measurements to energies above the lithium fluoride cutoff, new methods of producing circularly polarized light as well as sample handling will have to be developed. The various methods which may produce circularly polarized light of sufficient intensity at higher energies are discussed in other sections of this book as well as in the literature. [33,61-64] The future of natural and magnetic circular dichroism, as well as other techniques which use circularly polarized radiation, promises to be challenging and exciting as we go to higher energies.

Acknowledgement

This work was supported by the National Science Foundation under grants CHE 81-08534, DMR 77-21888, and CHE 77-08311, by the Research Corporation and by the Division of Sponsored Research at Florida Atlantic University. The assistance of the staff of the Synchrotron Radiation Center, University of Wisconsin-Madison is gratefully acknowledged.

References

1. L. Velluz, M. Legrand, and M. Grosjean,"Optical Circular Dichroism,"Academic Press, (1965).
2. J. C. Kemp, J. Opt. Soc. Am., 59, 950 (1969).
3. S. N. Jasperson and S. E. Schnatterly, Rev. Sci. Instrum., 40, 761 (1969).
4. M. Billardon and J. Badoz Compt. Rend. Paris, 262, 1672 (1966).
5. W. C. Johnson, Jr., Rev. Sci. Instrum., 35, 1375 (1964).
6. D. E. Eastman and J. J. Donnelon, Rev. Sci. Instrum., 41, 1648 (1970).
7. W. C. Johnson, Jr., Rev. Sci. Instrum., 42, 1283 (1971).
8. O. Schnepp, S. Allen, and E. F. Pearson, Rev. Sci. Instrum., 41, 1136 (1970).
9. S. Feinleib, and F. A. Bovey, Chem. Comm., 1968, 978 (1968).
10. K. P. Gross and O. Schnepp, Rev. Sci. Instr., 48, 362 (1977).

11. S. Allen and O. Schnepp, J. Chem. Phys., 59, 4547 (1973).
12. G. S. Pysh, Ann. Rev. Biophys. Bioengr., 5, 63 (1976).
13. A. Gedanken and M. Levy, Rev. Sci. Instr.,48, 1661 (1977).
14. S. Brahms, J. Brahms, G. Spach, and A. Brack, Proc. Natl. Acad. Sci. USA, 74, 3208 (1977).
15. A. F. Drake and S. F. Mason, Tetrahedron, 33, 104 (1977).
16. A. J. Duben and C. A. Bush, Anal. Chem., 52, 635 (1980).
17. J. D. Scott, W. S. Felps, G. L. Findley, and S. P. McGlynn, J. Chem. Phys., 68, 4673 (1978).
18. S. D. Allen, M. G. Mason, O. Schnepp, and P. J. Stephens, Chem. Phys. Lett., 30, 140 (1975).
19. W. C. Johnson, Jr., Ann. Rev. Phys. Chem., 29, 93 (1978).
20. S. F. Mason, "Molecular Optical Activity and the Chiral Discriminations," Cambridge University Press (1982).
21. S. Brahms and J. Brahms, J. Mol. Biol., 133, 149 (1980).
22. O. Schnepp in "Optical Activity and Chiral Discrimination," eds. S. F. Mason, D. Reidel (1979).
23. R. G. Nelson and W. C. Johnson, J. Am. Chem. Soc., 98, 4290 (1976).
24. J. Texter and E. S. Stevens, J. Chem. Phys., 70, 1440 (1979).
25. J. N. Liang and E. S. Stevens, Int. J. Biolog. Macromolecules, 14, 316 (1982).
26. A. J. Stipanovic and E. S. Stevens, Biopolymers, 20, 1565 (1981).
27. C. A. Bush, and A. Dua, S. Ralapati, C. D. Warren, G. Spik, G. Strecker, and J. Montreuil, J. Biolog. Chem., 257, 8199 (1982).
28. E. L. Bowman, M. Kellerman, and W. C. Johnson, Jr., Biopolymers, 22, 1045 (1983).
29. P. Manavala and W. C. Johnson, Jr., Nature, 305, 831 (1983).
30. Patricia Ann SNyder and W. Curtis Johsson, Jr., J. Am. Chem. Soc., 100, 2939 (1978).
31. A. Gedanken and O. Schnepp, Chem. Phys.., 12, 341 (1976).
32. Patricia Ann Snyder, Prudence M. Vipond and W. Curtis Johnson, Jr., Biopolymers, 12, 975 (1973).
33. Patricia Ann Snyder, Nucl. Instr. and Methods, 222, 364 (1984).
34. C. Pruett, Seya-Namioka Monochromator Description, Synchrotron Radiation Center, University of Wisconsin-Madison.
35. Patricia Ann Snyder and Ednor M. Rowe, VI International Conference on Vacuum Ultraviolet Radiation Physics Vol. III, 33, University of Virginia (1980).
36. Patricia Ann Snyder and Ednor M. Rowe, Nucl. Instr. and Meths., 172, 345 (1980).
37. G. C. Chen and J. T. Yang, Anal. Lett., 10, 1195 (1977)
38. P. A. Snyder, P. N. Lund, P. N. Schatz and E. M. Rowe, Chem. Phys. Lett., 82, 546 (1981).
39. E. H. Sharman, Ph.D. Thesis, University of Southern California, (1976).
40. P. A. Snyder, W. C. Johnson, Jr., A. Lin and E.M. Rowe, Thirty-Seventh Symposium on Molecular Spectroscopy, Ohio State University, 80, (1982).
41. P. J. Stephens, Ann. Rev. Phys. Chem., 25, 201 (1974).
42. P. J. Stephens, Adv. in Chem. Phys., 35, 197 (1976).
43. M. B. Robin, "Higher Excited States of Polyatomic Molecules," Vol. II, Academic Press, N.Y., 1975.
44. P. G. Wilkinson, Can. J. Phys., 24, 596 (1956).
45. P. G. Wilkinson, Can. J. Phys., 34, 643 (1956).

46. Patricia Ann Snyder, Paul N. Schatz and Ednor M. Rowe, Ann Israel
     Phys. Soc., 6, 144 (1984).
47. Patricia Ann Snyder, Paul N. Schatz, and Ednor M. Rowe, Chem.
     Phys. Lett., in press.
48. R. McDiarmid, J. Phys. Chem., 84, 64 (1980).
49. R. S. Mulliken, J. Chem. Phys., 66, 2443 (1977).
50. M. Brithlinder and S. D. Allen, Chem. Phys. Lett., 47, 32 (1977).
51. A. Gedanken, N. A. Kuebler and M. B. Robin, J. Chem. Phys., 76, 46
     (1982).
52. M. B. Robin, "Higher Excited States of Polyatomic Molecules,"
     Vol. I., Academic Press, N. Y., 1975.
53. A. Gedanken and O. Schnepp, Chem. Phys. Lett., 37, 373 (1976).
54. Patricia Ann Snyder, Michael Boyle, Paul N. Schatz, Roger C.
     Hansen, and Ednor M. Rowe, to be published.
55. P. R. Woodruff, C. H. Pruett, and I. H. Middleton, Nucl. Instr.
     and Meths., 172, 181 (1980).
56. J. Hormes, A. Klein, W. Krebs, W. Laaser, and J. Schiller, Nucl.
     Instr. and Meths., 208, 849 (1983).
57. J. C. Sutherland, P. C. Keck, K. P. Griffin and P. Z. Kakacs,
     Nucl. Instr. and Meths. 195, 375 (1982).
58. J. C. Sutherland, K. D. Griffin, P. C. Keck and P. Z. Takacs,
     Proc. Natl. Acad. Sci. USA, 78, 4801 (1981)
59. J. C. Sutherland, E. J. Desmond and P. Z. Takacs, Nucl. Instr. and
     Meths., 172, 195 (1980).
60. B. Alexa, M. A. Baig, J. P. Connerade, W. R. S. Gaiton, J. Hormes,
     T. A. Stavrakas, Nucl. Instr. and Meths. 208, 841 (1983).
61. P. D. Johnson and N. V. Smith, Nucl. Instr. and Meths., 222, 262
     (1984).
62. U. Heinzmann, I. Schafers, K. Thimm, A. Wolcke and J. Kessler, J.
     Phys., B12, 679 (1979).
63. Kwang Je Kim, Nucl. Instr. and Meths., 222, 11 (1984).
64. K. Halbach, Nucl. Instr. and Meths., 187, 109 (1981).

MEASURING THE MUELLER MATRIX BY
A MULTIMODULATOR SCATTERING INSTRUMENT

Cathy D. Newman,
John H. May,
    and
Fritz S. Allen

Department of Chemistry
University of New Mexico
Albuquerque, New Mexico 87131

ABSTRACT

An instrument is being constructed to measure the Mueller matrix for a scattering sample. The instrument is a modified version of that described by Fry, et al. A summary of the theory is given with pertinent extensions. A method for this precise calibration of photoelastic modulators is described.

INTRODUCTION

We present here a review of an instrument which is presently under construction in our laboratory. In principle, this device could be extended to much shorter wave lengths, including those accessible through the synchrotron. The instrument is being constructed to measure the scattering or Mueller matrix for a sample material. The scattering matrix contains all the information relating to the linear optical effects of the sample. At zero angle, this includes absorption, linear and circular dichroism, linear and circular birefringence, and other complex polarization exchange phenomena. At angles other than zero, the matrix gives the magnitude of the Raleigh scattering, the differential linear and circular scattering, and terms describing the interaction of various polarization forms in the scattering process. (For example, the extent of circular scattering as a consequence of linearly polarized incident radiation, etc.) Clearly, knowledge of the scattering matrix as

a function of angle and wavelength is of great value for any
sample material.

Previous investigators have been interested in the
determination of the scattering matrix.[1-3] Our method is
based primarily on the technology of Thompson and Fry.[4]

Over a decade ago Hunt and Huffman[5] utilized a single
photoelastic modulator as designed by Kemp[6] to modulate the
polarization of the light on the scattering cell. Thompson and
Fry generalized the methods of Hunt and Huffman and constructed
an instrument which has four modulators in the optical train.
This instrument provided for the first direct simultaneous
measurement of all sixteen elements of the scattering matrix.
To fully describe the interaction of light with a scattering
medium, we must take into account all the effects due to the
polarization of the light[7]. Formally, then the problem
reduces to obtaining both a complete description of the
intensity and polarization of the radiation as well and as a
characterization of the response of the particle to the incident
radiation.

A formulation due to G. G. Stokes[8] is used to represent an
arbitrarily polarized beam of light. This formulation uses four
parameters to describe the radiant intensity, the degree of
polarization, both linear and circular, and the azimuth of the
linear polarization:

$$I = <E_{,,}^2 + E_{,}^2>$$

$$Q = <E_{,,}^2 - E_{,}^2>$$

$$U = <2E_{,,}E_{,}\cos(\delta)>$$

$$V = <2E_{,,}E_{,}\sin(\delta)>$$

(1)

where the angular brackets represent time average, $E_{,,}$ and $E_{,}$
are the parallel and perpendicular components of the scattered
electric field with respect to the scattering plane and δ is
the phase shift between the parallel and perpendicular
components. I represents the total radiant intensity; Q the
excess in intensity of light transmitted by a polarizer which
accepts linear polarization in the parallel direction over a
polarizer which accepts linear polarization along the
prependicular direction. U has a meaning equivalent to Q, but
with the polarizer oriented at 45° and 135° from the
perpendicular direction, and V, describes the excess in

intensity of light transmitted by a device that accepts right
circular polarization as opposed to left. These four components
constitute a column vector called the Stokes vector. When light
described by such a vector is incident upon a sample, its
direction, intensity and state of polarization is changed
depending on the optical properties and hence the structure of
the scattering particles. That is, a given sample material may
alter the total intensity and linear and circular polarization
characteristics of the light upon transmission. If one neglects
nonlinear effects, it is clear that the initial and emergent
Stokes vectors must be coupled by a 4x4 matrix so that:

$$\vec{S}_e = M \vec{S}_i \tag{2}$$

where S_e is the emergent Stokes vector, S_i the initial
Stokes vector, and M is the coupling or scattering matrix. The
elements of this scattering matrix will be a function of the
properties of the scatterer and the scattering direction.

In general, any optical element that changes the intensity
and/or direction and/or state of polarization of the incident
radiation can be represented by an equivalent matrix. Such
matrices are called Mueller matrices. This linear formulation
of the interaction of optical beams and matter is known as the
Mueller calculus.

It is evident that by using at least four different incident
polarizations, it would be possible to invert Equation 2 and
obtain the sixteen elements of the scattering matrix. Other
discussions in this volume have defined the physical meaning of
many of the terms in this scattering matrix. Let us consider
the Mueller calculus for a series of optical elements.
Successive applications of Equation 2 lead us to the conclusion
that it is possible to obtain the cumulative effect on the
optical beam of a series of elements as a simple matrix product
of the Mueller matrices for the serial optical elements. This
results in Equation 3.

$$\vec{S}_e = \cdots \underset{\sim 3}{M} (\underset{\sim 2}{M} (\underset{\sim 1}{M} \bullet \vec{S}_i)) = \cdots \underset{\sim 3}{M}\underset{\sim 2}{M}\underset{\sim 1}{M} \vec{S}_i \tag{3}$$

By this means it is easy to describe even a very complex optical
beam provided that the Mueller matrices are available for all
the elements.

In Figure 1, we show a schematic diagram of the optical
components of the instrument under construction. MOD 1,2,3 and
4 represent the polarization modulators. In our device, the
angle between the principal axis of each modulator and the
reference plane is denoted by γ_i ($i = 1,2,3,4$). POL 1 and
POL 2 are linear polarizers with transmission axis parallel to
the scattering plane. The sample in this figure is denoted by F.

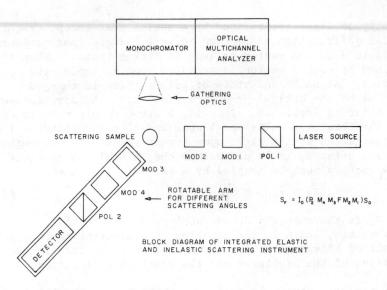

FIGURE 1:   Schematic Instrumental Diagram

The Mueller matrix for polarizers 1 and 2 is then:

$$P_{1 \text{ or } 2} = 1/2 \begin{bmatrix} 1 & -1 & 0 & 0 \\ -1 & 1 & 0 & 0 \\ 0 & 0 & 0 & 0 \\ 0 & 0 & 0 & 0 \end{bmatrix} \tag{4}$$

the matrix for the modulator 1 and 4 with $\gamma = \pi/4$

$$M(1 \text{ or } 4) = \begin{bmatrix} 1 & 0 & 0 & 0 \\ 0 & \cos\delta & 0 & \sin\delta \\ & (\delta 1 \text{ or } \delta 4) & & (\delta 1 \text{ or } \delta 4) \\ 0 & 0 & 1 & 0 \\ 0 & -\sin\delta & 0 & \cos\delta \\ & (\delta 1 \text{ or } \delta 4) & & (\delta 1 \text{ or } \delta 4) \end{bmatrix} \tag{5}$$

for modulators 2 and 3, with $\gamma = 0$ it is:

$$M(2 \text{ or } 3) = \begin{bmatrix} 1 & 0 & 0 & 0 \\ 0 & 1 & 0 & 0 \\ 0 & 0 & \cos\delta & -\sin\delta \\ & & (\delta 2 \text{ or } \delta 3) & (\delta 2 \text{ or } \delta 3) \\ 0 & 0 & \sin\delta & \cos\delta \\ & & (\delta 2 \text{ or } \delta 3) & (\delta 2 \text{ or } \delta 3) \end{bmatrix} \tag{6}$$

where δ_i (i = 1,2,3,4) is the retardance of the ith modulator. Notice that the 1st polarizer ensures that the initial state of polarization such that

$$S_i = I_i \begin{bmatrix} 1 \\ 1 \\ 0 \\ 0 \end{bmatrix} \qquad (7)$$

where the intensity has been factored out as I_i. Similarly, the 2nd polarizer ensures that the light incident on the photomultiplier tube is

$$S_e = I_e \begin{bmatrix} 1 \\ 1 \\ 0 \\ 0 \end{bmatrix} \qquad (8)$$

Notice that the polarization of the light incident in the photomultiplier tube is constant. Thus, polarization dependence on the sensitivity of the photomultiplier tube cannot lead to systematic errors.

Notice that because of (8), we need only compute the I_e that results when the initial Stoke's parameter S_e is operated on by all the elements in the optical train and by its interaction with the sample, i.e.:

$$\bar{S}_f = I_o (P_2 M_4 M_3 F M_2 M_1) \bar{S}_o \qquad (9)$$

After performing the products indicated in (9) with the help of equations (4) to (8) and using the notation f_{ij} for the elements of the scattering matrix of the sample, we obtain:[4]

$$I_f = \frac{I_o}{2} (f_{11} + f_{12} \cos\delta_1 + f_{13} \sin\delta_1 \sin\delta_2 \qquad (10)$$

$$- f_{14} \sin\delta_1 \cos\delta_2 + f_{21} \cos\delta_4 + f_{22} \cos\delta_1 \cos\delta_4$$
$$+ f_{23} \sin\delta_1 \sin\delta_2 \cos\delta_4 - f_{24} \sin\delta_1 \cos\delta_2 \cos\delta_4$$
$$+ f_{31} \sin\delta_3 \sin\delta_4 + f_{32} \cos\delta_1 \sin\delta_3 \sin\delta_4$$
$$+ f_{33} \sin\delta_1 \sin\delta_2 \sin\delta_3 \sin\delta_4$$
$$- f_{34} \sin\delta_1 \cos\delta_2 \sin\delta_3 \sin\delta_4$$
$$+ f_{41} \cos\delta_3 \sin\delta_4 + f_{42} \cos\delta_1 \cos\delta_3 \sin\delta_4$$
$$+ f_{43} \sin\delta_1 \sin\delta_2 \cos\delta_3 \sin\delta_4$$
$$- f_{44} \sin\delta_1 \cos\delta_2 \cos\delta_3 \sin\delta_4)$$

Each of the elements of the sample scattering matrix appear here
multiplied by a unique combination of trigonometric functions of
the retardances of the modulators. Now, since the modulator
retardances can be varied sinusoidally at different frequencies,
we then write:

$$\delta_i = \delta_{oi} \cos\omega_i t \qquad (11)$$

where δ_{oi} is the amplitude of the retardance of the ith
modulator.

With this form for the retardance, the functions $\sin\delta_i$
and $\cos\delta i$ can be expanded in terms of Bessel functions of
the retardation amplitudes[29],

$$\cos\delta_i = J_0(\delta_{oi}) - 2J_2(\delta_{oi})\cos2\omega_i t + 2J_4(\delta_{oi})\cos4\omega_i t - \ldots (12)$$

$$\sin\delta_i = 2J_1(\delta_{oi})\cos\omega_i t - 2J_3(\delta_{oi})\cos3\omega_i t + \ldots \qquad (13)$$

These results are substituted into Eq. (10), and all products
are expanded to give the Fourier components. We find that the
intensity I_0 has a dc component plus sinusoidal terms with
amplitudes proportional to products of the Bessel functions and
the matrix elements f_{ij}. The frequency of every Fourier
component is given by

$$k\omega_1 \pm l\omega_2 \pm m\omega_3 \pm n\omega_4, \quad k,l,m,n = 0,1,2,3, \ldots \qquad (14)$$

Following Fry[4] we give the expansion of the coefficient of
f_{14} in Eq. (11).

$$\begin{aligned}
\sin\delta_1\cos\delta_2 = \ & 2J_1(\delta_{o1})J_0(\delta_{o2})\cos\omega_1 t \\
& - 2J_3(\delta_{o1})J_0(\delta_{o2})\cos3\omega_1 t \\
& - 2J_1(\delta_{o1})J_2(\delta_{o2}) \qquad (15) \\
& \times \{\cos(\omega_1 + 2\omega_2)t + \cos(\omega_1 - 2\omega_2)t\} \\
& + 2J_3(\delta_{o1})J_2(\delta_{o2}) \\
& \times \{\cos(3\omega_1 + 2\omega_2)t \\
& + \cos(3\omega_1 - 2\omega_2)t\} + \ldots
\end{aligned}$$

Thus, the frequencies at which f_{14} appears in the intensity
are given by Eq. (14) with $m = n = 0$, k = odd integers, and l =
even integers and zero. The linear combinations of frequencies
at which the more difficult matrix elements appear in the
Fourier expansion of the intensity are given in Table 1.

Table 1

Amplitude and Frequency Characteristic as Multiples of $\omega_1 \ldots \omega_4$

Element	Amplitude	Frequency Multiple	Range
f_{31}	$I_0 J_{2n+1} J_{2m+1}(-1)^{n+m}$	$0\omega_1 \pm 0\omega_2 \pm (2n+1)\omega_3 \pm (2m+1)\omega_4$	$n,m=0\to+\infty$
f_{32}	$I_0 J_{2n} J_{2m+1} J_{2p+1}(-1)^{n+m+p}$	$2n\omega_1 \pm 0\omega_2 \pm (2m+1)\omega_3 \pm (2p+1)\omega_4$	$n=1\to+\infty$ $m,p=0\to+\infty$
f_{33}	$I_0 J_{2n+1} J_{2m+1} J_{2p+1}$ $J_{2q+1}(-1)^{n+m+p+q}$	$(2n+1)\omega_1 \pm(2m+1)\omega_2 \pm(2p+1)\omega_3 \pm(2q+1)\omega_4$	$n,m=0\to+\infty$ $p,q=0\to+\infty$
f_{34}	$I_0 J_{2n+1} J_{2m} J_{2p+1}$ $J_{2q+1}(-1)^{n+m+p+q}$	$(2n+1)\omega_1 \pm 2m\omega_2 \pm(2p+1)\omega_3 \pm(2q+1)\omega_4$	$m=1\to+\infty$ $n,p,q=-\infty$
f_{41}	$I_0 J_{2n} J_{2m+1}(-1)^{n+m}$	$0\omega_1 \pm 0\omega_2 \pm 2n\omega_3 \pm (2m+1)\omega_4$	$n=1\to+\infty$ $m=0\to+\infty$
f_{42}	$I_0 J_{2n} J_{2m} J_{2p+1}(-1)^{n+m+p}$	$2n\omega_1 \pm 0\omega_2 \pm 2m\omega_3 \pm(2p+1)\omega_4$	$n,m=1\to+\infty$ $m=0\to+\infty$
f_{43}	$I_0 J_{2n+1} J_{2m+1} J_{2p}$ $J_{2q+1}(-1)^{n+m+p+q}$	$(2n+1)\omega_1 \pm(2m+1)\omega_2 \pm 2p\omega_3 \pm(2q+1)\omega_4$	$n,m,q=0\to+\infty$ $p=1\to+\infty$
f_{44}	$I_0 J_{2n+1} J_{2m} J_{2p}$ $J_{2q+1}(-1)^{n+m+p+q}$	$(2n+1)\omega_1 \pm 2m\omega_2 \pm 2p\omega_3 \pm(2q+1)\omega_4$	$n,q=0\to+\infty$ $m,p=1\to+\infty$

To ensure the distinct measurement of each matrix element, we will require that there be a unique reference frequency in the Fourier spectrum of Eq. (10) corresponding to each matrix element. Experimental considerations make it desirable that the selected component be one of the leading terms in the Fourier expansion and that the frequency be compatible with standard phase-sensitive detection techniques (< 200 kHz).

We can freely choose the four primary frequencies ω_i as well as the amplitudes of the retardances δ_{oi}, so as to satisfy these requirements.

By taking $\delta_{oi} = 2.404$ rad then : (16)

$$J_0(\delta_{oi}) = 0$$

Each of the terms in the matrix can be separated from the rest:

First the expansion in (12) and (13) must be carried out to high enough terms so that the approximation in the truncation of the series is negligible with respect to the value of the elements in the matrix that are to be measured. Such extended expressions tend to mix additional frequency combinations in

each term and therefore some degeneracy between frequencies for
the different matrix elements may appear. The combinations of
matrix elements will however be known for each frequency and a
system of linear equations can be written and solved to
determine the true matrix elements.

It is clear that by setting the retardance of the modulators
to 2.404 radians a major simplification of the expansions
carried out above is achieved. However, this makes calibration
of the modulators a difficult problem. The response of an
optical system when the modulator is calibrated to 2.404 radians
is not an obvious and easy-to-characterize function. However,
in another part of this volume[9] we have described a means to
model the intensity for a modulator placed between crossed
polarizers. This methodology can be employed to determine what
the optical signal will look like when the modulator is
operating to give a retardance of 2.404 radians. Knowing the
waveform from these calculations for the optical signal, one can
proceed to increase the modulation power until the desired
waveform is achieved. In general, this is a difficult judgement
to make with precision. Instead, we have developed another
procedure which makes calibration to any arbitrary degree of
retardance a straightforward matter.

The method proceeds as follows. The techniques described
earlier are employed to determine the waveform for the desired
degree of retardance. This waveform is Fourier transformed to
obtain the contributions at the fundamental and higher harmonics
of the crystal. The actual optical signal is spectrum analyzed,
and the calculated Fourier efficients are compared with the
experimental pattern. The peak heights in the spectrum analyzer
are very sensitive to the power setting of the modulator. Very
small changes in this power setting significantly affect the
ratios of the peak heights. As a consequence it becomes much
easier to determine the exact wave shape in the frequency domain
than in an actual examination of the waveform itself.

As an example of this procedure, we show in Figure 2 two
waveforms along with the spectrum analysis pattern for each of
the waveforms. The waveforms and spectral data were obtained at
significantly different power settings on the modulator.
Nonetheless, the waveforms are very similar but the spectral
patterns are significantly different. In general, it is
possible to calibrate a modulator with arbitrary retardation to
1 part in 1000 on a reproducible basis. Since the polarization
state of the radiation in our instrumental system is extremely
important, it is necessary to use these techniques to calibrate
the modulators in order to provide reliable measurements of the
scattering matrix.

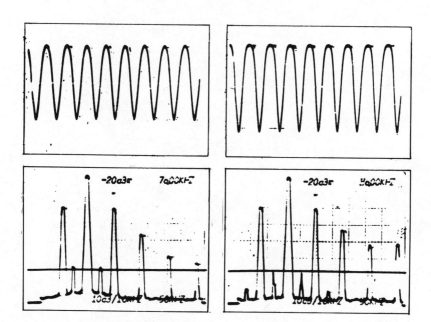

FIGURE 2.   Modulator Waveforms in Time and Frequency Domains

REFERENCES

1. G. F. Beardsley, Jr., J. Opt. Soc. Am., 58, 52 (1968).

2. A. C. Holland and G. Gagne, Appl. Opt., 9, 1113 (1970).

3. G. V. Rozenberg, Yu. S. Lyubovtseva, Ye. A. Kadyshevich, and N. R. Amnuil, Atmospheric and Oceanic Physics, 6, 747 (1970).

4. R. C. Thomson, J. R. Bottiger and E. S. Fry, Appl. Opt., 19, 1323-1332 (1980).

5. A. J. Hunt and D. R. Huffman, Rev. Sci. Instrum., 44, 1753 (1973).

6. J. C. Kemp, J. Opt. Soc. Am., 59, 950 (1969).

7. R. C. Thomson, Ph.D. Thesis, Texas A & M University, (1978).

8. G. G. Stokes, Trans. Camb. Phil. Soc., 9, 399 (1852).

9. William H. Rahe, Robert J. Fraatz, Fritz S. Allen, "High Speed Photoelastic Modulation", refer to this volume, 109.

THE MUELLER SCATTERING MATRIX ELEMENTS

FOR RAYLEIGH SPHERES

William S. Bickel

Physics Department
University of Arizona
Tucson, Arizona

Abstract

All 16 elements of the Mueller light scattering
matrix for a Rayleigh sphere are predicted from systematic
application of the input-output polarizers used to measure
scattering intensities from a classical electric dipole.

INTRODUCTION

Experimentalists and theorists often discuss light scattering
results in the context of Stokes vectors and Mueller matrices. A
particular light scattering data point is generally nothing more than
the value of a particular matrix element measured at a particular
angle. Of the infinite number of angles and polarization combinations
to choose from, only a small number of highly motivated measurements
is needed to completely characterize the scatterer. The vector-matrix
representation tells exactly what they are. Because matrix methods are
so powerful in describing the light scattering interaction, it is
important to understand their role as "the data" from light-scattering
experiments.

We will begin by describing the most general light-scattering
experiment and derive the general Mueller scattering matrix elements
S_{ij} in terms of the input-output Stokes vectors required to
characterize the scattering process. Finally, we will use the results
to predict the four nonzero Mueller matrix elements for small Rayleigh
sphere (electric dipole radiator).

The experimental setup that can measure the polarized intensities
scattered by the scatterer [S] into the angles θ and ϕ is shown in Fig.
1. The input optics can be selected to be: an open hole [0], or
horizontal linear polarizer |h|, or +45 linear polarizer [+], or right-
hand circular polarizer [r]. The exit optics choices are the same and
can be chosen independently of the input optics. In addition, they can
be swung with the detector through the scattering angle θ from 0° to
180°.

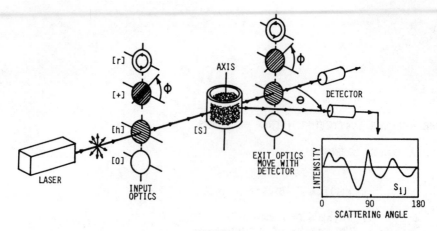

FIGURE 1

The following example shows how to determine what scattering matrix elements S_{ij} are involved when a particular set of input-output polarizers are used to prepare and analyze the scattered light. We assume that the arbitrary scatterer [S] is illuminated with horizontally polarized light |h|. The scattered Stokes vector will be $|V_s| = [S]*|h|$. In terms of the specific Mueller matrices and Stokes vectors involved we have

$$
\begin{array}{ccccc}
[S] & * & |h| & = & |V_s| \\
\begin{bmatrix} S_{11} & S_{12} & S_{13} & S_{14} \\ S_{21} & S_{22} & S_{23} & S_{24} \\ S_{31} & S_{32} & S_{33} & S_{34} \\ S_{41} & S_{42} & S_{43} & S_{44} \end{bmatrix} & & \begin{vmatrix} 1 \\ 1 \\ 0 \\ 0 \end{vmatrix} & = & \begin{vmatrix} S_{11} + S_{12} \\ S_{12} + S_{22} \\ S_{13} + S_{32} \\ S_{14} + S_{42} \end{vmatrix}
\end{array}
$$

We see that the scatterer [S] mixes the initially pure polarization state |h| to produce a scattered Stokes vector with mixed polarizations. In addition, each Stokes component is now a mixture of two matrix elements. The first component $(S_{11} + S_{12})$ is the total intensity.

If this scattered light is now passed through a +45 linear polarizer [+], we get $[+]*|V_s| = |V_f|$, which will be detected by the detector. Specifically we have:

$$
\begin{array}{cccc}
[+] & * & |V_s| & |V_f| \\
\begin{bmatrix} 1 & 0 & 1 & 0 \\ 0 & 0 & 0 & 0 \\ 1 & 0 & 1 & 0 \\ 0 & 0 & 0 & 0 \end{bmatrix} & & \begin{vmatrix} S_{11} + S_{12} \\ S_{21} + S_{22} \\ S_{31} + S_{32} \\ S_{41} + S_{42} \end{vmatrix} & = \begin{vmatrix} S_{11} + S_{12} + S_{31} + S_{32} \\ 0 \\ S_{11} + S_{12} + S_{31} + S_{32} \\ 0 \end{vmatrix}
\end{array}
$$

The first component of the final Stokes vector, is now a mixture of four matrix elements. The first element sum, $(S_{11} + S_{12} + S_{31} + S_{32})$, has special experimental significance, since it is the total intensity that will be measured by the detector.

We put the results of all such calculations for all 16 Stokes vector combinations into a final matrix array shown in Fig. 2. Each matrix element lable S_{ij} is in the uppermost left-hand corner of each

S_{11}	S_{12}	S_{13}	S_{14}
S_{11}	$S_{11}+S_{12}$ $S_{11}-S_{12}$	$S_{11}+S_{13}$ $S_{11}-S_{13}$	$S_{11}+S_{14}$ $S_{11}-S_{14}$
S_{21}	S_{22}	S_{23}	S_{24}
$S_{11}+S_{21}$ $S_{11}-S_{21}$	$S_{11}+S_{12}+S_{21}+S_{22}$ $S_{11}+S_{12}-S_{21}-S_{22}$ $S_{11}-S_{12}+S_{21}-S_{22}$ $S_{11}-S_{12}-S_{21}+S_{22}$	$S_{11}+S_{13}+S_{21}+S_{23}$ $S_{11}+S_{13}-S_{21}-S_{23}$ $S_{11}-S_{13}+S_{21}-S_{23}$ $S_{11}-S_{13}-S_{21}+S_{23}$	$S_{11}+S_{14}+S_{21}+S_{24}$ $S_{11}+S_{14}-S_{21}-S_{24}$ $S_{11}-S_{14}+S_{21}-S_{24}$ $S_{11}-S_{14}-S_{21}+S_{24}$
S_{31}	S_{32}	S_{33}	S_{34}
$S_{11}+S_{31}$ $S_{11}-S_{31}$	$S_{11}+S_{12}+S_{31}+S_{32}$ $S_{11}+S_{12}-S_{31}-S_{32}$ $S_{11}-S_{12}+S_{31}-S_{32}$ $S_{11}-S_{12}-S_{31}+S_{32}$	$S_{11}+S_{13}+S_{31}+S_{33}$ $S_{11}+S_{13}-S_{31}-S_{33}$ $S_{11}-S_{13}+S_{31}-S_{33}$ $S_{11}-S_{13}-S_{31}+S_{33}$	$S_{11}+S_{14}+S_{31}+S_{34}$ $S_{11}+S_{14}-S_{31}-S_{34}$ $S_{11}-S_{14}+S_{31}-S_{34}$ $S_{11}-S_{14}-S_{31}+S_{34}$
S_{41}	S_{42}	S_{43}	S_{44}
$S_{11}+S_{41}$ $S_{11}-S_{41}$	$S_{11}+S_{12}+S_{41}+S_{42}$ $S_{11}+S_{12}-S_{41}-S_{42}$ $S_{11}-S_{12}+S_{41}-S_{42}$ $S_{11}-S_{12}-S_{41}+S_{42}$	$S_{11}+S_{13}+S_{41}+S_{43}$ $S_{11}+S_{13}-S_{41}-S_{43}$ $S_{11}-S_{13}+S_{41}-S_{43}$ $S_{11}-S_{13}-S_{41}+S_{43}$	$S_{11}+S_{14}+S_{41}+S_{44}$ $S_{11}+S_{14}-S_{41}-S_{44}$ $S_{11}-S_{14}+S_{41}-S_{44}$ $S_{11}-S_{14}-S_{41}+S_{44}$

FIGURE 2

matrix element block. The symbols to the immediate right of S_{ij} represent the kind of light involved in the measurement. Symbols below the dotted line in each box show the complementary orientations of the input-output polarizations. The actual matrix element combinations involved in that intensity measurement are given on the right of each symbol pair. Each matrix element sum is the intensity measured using a particular input-output polarization combination. S_{11} is determined with one measurement. The matrix elements of row 1, column 1 need two measurements, while all the others need four. Therefore 49 θ-dependent intensity measurements are needed to uniquely determine the 16 θ-dependent matrix elements which completely characterize the scatterer. They are displayed in Fig. 3.

A previous paper (Ref. 1) calculated the functional relationships for the various matrix element sums (intensities) that occur when the "scatterer" is a non-scattering perfect linear polarizer, circular

S_{11} ✳ ✳	S_{12} ↔ ✳	S_{13} ✎ ✳	S_{14} ◯ ✳
I_{00}	$I_{HO} - I_{VO}$	$I_{+0} - I_{-0}$	$I_{LO} - I_{RO}$
S_{21} ✳ ↔	S_{22} ↔ ↔	S_{23} ✎ ↔	S_{24} ◯ ↔
$I_{OH} - I_{OV}$	$(I_{HH} + I_{VV}) - (I_{VH} + I_{HV})$	$(I_{+H} + I_{-V}) - (I_{-H} + I_{+V})$	$(I_{LH} + I_{RV}) - (I_{RH} + I_{LV})$
S_{31} ✳ ✎	S_{32} ↔ ✎	S_{33} ✎ ✎	S_{34} ◯ ✎
$I_{0+} - I_{0-}$	$(I_{H+} + I_{V-}) - (I_{V+} + I_{H-})$	$(I_{++} + I_{--}) - (I_{-+} + I_{+-})$	$(I_{L+} + I_{R-}) - (I_{R+} + I_{L-})$
S_{41} ✳ ◯	S_{42} ↔ ◯	S_{43} ✎ ◯	S_{44} ◯ ◯
$I_{0L} - I_{0R}$	$(I_{HL} + I_{VR}) - (I_{VL} + I_{HR})$	$(I_{+L} + I_{-R}) - (I_{-L} + I_{+R})$	$(I_{LL} + I_{RR}) - (I_{RL} + I_{LR})$

FIGURE 3

polarizer, or quarter wave plate. These results calibrate the scattering instrument and establish a "frame of reference" for a real scatterer which will behave in part like a linear polarizer, quarter wave plate, circular polarizer, etc., and perhaps like mixtures of them.

SCATTERERS AND THE SCATTERING MATRIX

Scatterers [S] can be divided into three categories: (1) small Rayleigh spheres, (2) large Mie spheres, and (3) polydispersed, nonspherical, irregular, random-oriented particulates. The scattering intensities and polarizations from the first two perfect sphere systems can be calculated exactly from Maxwell's equations and electromagnetic theory. This paper shows how to predict the experimentally measured matrix elements for a small Rayleigh sphere using only knowledge about the electric fields and intensity distributions for a driven dipole radiator and the effect of polarizers on the incident and scattered light. This procedure gives insight into the scattering process and demonstrates how polarized intensity measurements are used to calculate matrix elements. The Rayleigh curves are fundamental, easy to calculate, and represent the starting point for studies of larger and irregular particles. The following procedure uses the optical setup of Fig. 1 to make the measurements described in Figs. 2 and 3.

THE MATRIX ELEMENT S_{11}

Matrix element S_{11} is determined from the single measurement of the θ-dependent total intensity scattered from a scatterer illuminated with unpolarized light. The exact intensity function for a Rayleigh particle is $I = k(1 + \cos^2\theta)$, which in this special case is independent of frequency. Figure 4 shows how this function would by measured by the optical system shown in Fig 1. Note that as the open hole and detector scan from $0°$ to $180°$, the measured intensity $I(oo)$, as plotted on a strip chart, will be exactly $I = k(1 + \cos^2\theta)$. The small intensity dip at $90°$ shows that a small Rayleigh sphere scatters unpolarized light almost isotropically.

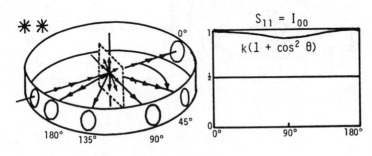

FIGURE 4

THE MATRIX ELEMENT S_{12}

Matrix element S_{12} is determined from two measurements. The total scattered intensity must be measured for the scatterer, illuminated first with a horizontal and then with a vertical linear polarization. Note from the two optical arrangements shown in Fig. 5 that as the open hole and detector are scanned $180°$ for horizontally polarized illumination, the intensity measured is $I = I(o)1/2(1 + \cos2\theta)$ $= (S_{11} + S_{12})$. When they are scanned for vertically polarized illumination the intensity is $I = I(0) = (S_{11} - S_{12})$. From Figs. 2 and 3 we see that $2 S_{12} = [(S_{11} + S_{12}) - (S_{11} - S_{12})]$, which is the same as $I(vo) - I(ho)$. The resultant "intensity" curve for S_{12}, therefore varies as $-\sin^2\theta$. This curve is the horizontal-vertical polarization function for dipole radiation.

THE MATRIX ELEMENT S_{33}

Matrix element S_{33} is determined from four measurements, as indicated in Figs. 2 and 3. Figure 6 shows the optical arrangement of the four polarization combinations needed for the measurements described below.

A. Measure $I(++) = S_{11} + S_{13} + S_{31} + S_{33}$ (+45 goes to + 45)
B. Measure $I(+-) = S_{11} + S_{13} - S_{31} - S_{33}$ (+45 goes to - 45)
   Then Subtract $I(++) - I(+-) = 2(S_{31} + S_{33})$

C. Measure $I(-+) = S_{11} - S_{13} + S_{31} - S_{33}$ (-45 goes to +45)
D. Measure $I(--) = S_{11} - S_{13} - S_{31} + S_{33}$ (-45 goes to -45)
   Then Subtract $I(-+) - I(--) = 2(S_{31} - S_{33})$

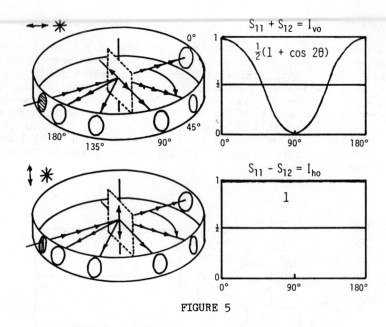

FIGURE 5

Finally, compute $[I(++) - I(+-)] - [I(-+) - I(--)] =$
                    $[I(++) + I(--)] - [I(-+) + I(+-)] = 4S_{33}$

The resultant "intensity" curve for S_{33} therefore varies as $\cos\theta$. This curve is related to the $+45 - -45$ polarization function for dipole radiation.

THE MATRIX ELEMENT S_{34}

Matrix element S_{34}, involving circularly polarized light, is determined by the four measurements indicated in Figs. 2 and 3. Applying the same procedure described for S_{33} shows that S_{34} for small Rayleigh particles is zero. By definition they are too small to evoke any geometrical or optical path difference between any extreme rays that they scatter. Larger spheres have a non-zero S_{34} matrix element.

ALL OTHER MATRIX ELEMENTS S_{ij}

All other matrix elements for Rayleigh spheres are either zero or identical to ones just calculated. This can be easily shown by applying the same procedures described above.

THE LIGHT SCATTERING MATRIX FOR SPHERES

We conclude our discussion with some comments about the light scattering matrix for spheres in general, regardless of refractive index and absorption. These comments are valid for individual sphere mixtures and polydispersed systems.

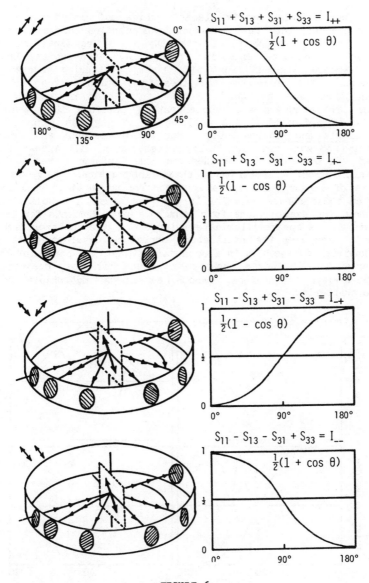

$$S_{11} + S_{13} + S_{31} + S_{33} = I_{++}$$

$$\tfrac{1}{2}(1 + \cos \theta)$$

$$S_{11} + S_{13} - S_{31} - S_{33} = I_{+-}$$

$$\tfrac{1}{2}(1 - \cos \theta)$$

$$S_{11} - S_{13} + S_{31} - S_{33} = I_{-+}$$

$$\tfrac{1}{2}(1 - \cos \theta)$$

$$S_{11} - S_{13} - S_{31} + S_{33} = I_{--}$$

$$\tfrac{1}{2}(1 + \cos \theta)$$

FIGURE 6

1. For all spheres, $S_{11} = S_{22}$, $S_{12} = S_{21}$, $S_{33} = S_{44}$, and $S_{34} = -S_{43}$ with all other matrix elements $S_{ij} = 0$.

2. For all spheres, S_{12} and S_{34} are always zero at $= 0°$ and $180°$. S_{33} is 100% at $0°$ and -100% at $180°$. These bounds hold regardless of how the S_{ij} curve may fluctuate between $0°$ and $180°$.

3. For Rayleigh spheres, S_{11} and S_{12} are symmetric while S_{33} is antisymmetric about $0 = 90°$. S_{34} is zero everywhere.

The four Rayleigh matrix element curves calculated above can be converted directly into the well-known polarization curves by dividing S_{12}, S_{33}, and S_{34} by the total intensity S_{11}. The final curves are summarized by the dotted lines in Fig. 7. They represent the "starting point reference curves" for all light scattering curves from particulates. They form the baselines from which curves for larger particles grow and to which curves from complex systems might approach. These curves are independent of particle size, shape, and orientation for individual and collections of Rayleigh particles that scatter single and independently. For non-independent and multiple scattering from even small particles, these curves will appear slightly distorted. It is proper to consider that the Rayleigh curves contain no phase information. Oscillatory phase information, which is truly indicative of larger particles and responsive to particle size changes, appears on these curves only for larger particles. The solid lines of Fig. 7 show, for example, the four nonzero matrix elements for larger r = 0.30 micron Mie sphere illuminated with 0.4416 micron laser light. It is easy to see how the oscillatory structure developed out of the smooth Rayleigh curves. It is also apparent that the phase information characteristic of larger sphere systems can be destroyed by polydispersivity. Therefore, smooth curves do not imply Rayleigh particle scattering.

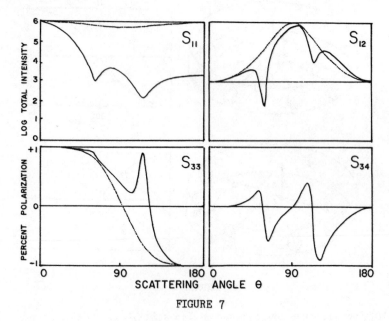

FIGURE 7

ACKNOWLEDGMENTS

The author acknowledges research support from the Army Chemical Systems Laboratory.

REFERENCE

1. William S. Bickel and Wilbur M. Bailey, Stokes Vectors, Mueller Matrices and Polarized Light Scattering, Am. J. Phys. (Jan. 1985), to be published.

CIRCULAR INTENSITY DIFFERENTIAL SCATTERING MEASUREMENTS OF PLANAR AND FOCAL CONIC ORIENTATIONS OF CHOLESTERIC LIQUID CRYSTALS

K. Hall[¶], K. S. Wells[*], D. Keller[*], B. Samori[¶¶],
M. F. Maestre[+], I. Tinoco, Jr.[¶], and C. Bustamante[*]

[¶] Department of Chemistry
University of California, Berkeley

[*] Department of Chemistry
University of New Mexico, Albuquerque

[¶¶] Istituto di Chimica degli Intermedi
Universitá di Bologna

[+] Lawrence Berkeley Laboratory
Division of Biology and Medical Physics

ABSTRACT

We have applied the recently developed technique of circular intensity differential scattering (CIDS) to the study of oriented liquid crystals. The chirality of the liquid crystals and the ease of manipulation of their helical parameters make them an ideal system for investigating the dependence of the CIDS pattern on handedness and pitch. We have studied both right- and left-handed liquid crystals of pitch from 460 nm to 4 microns, with the helix axis oriented either parallel (planar orientation) or perpendicular (focal conic orientation) to the incident beam. The results showed that CIDS is sensitive to the handedness of the helix, for when two liquid crystals of the same pitch but opposite handedness were compared, the signs of their respective CIDS patterns were reversed.

Difficulties encountered in performing these experiments are discussed. The comparison of the results with a theoretical model that uses the second Born approximation shows good agreement.

1. INTRODUCTION

Circular intensity differential scattering (CIDS) is a technique that is sensitive to the chiral geometry of an object. For a helical object, these geometric parameters are pitch, radius, and handedness. The theory has been developed for both oriented and randomly oriented systems[1-4] and experimental results have been reported for rotationally averaged systems of helical octopus sperm[5] and bacteriophages[6]. Here, we describe the scattering of an oriented system, using cholesteric liquid crystals.

Cholesteric liquid crystals form macroscopic helices that can be manipulated easily to alter the pitch. These mesophases can be obtained by dissolving a chiral molecule in a nematic mesophase. These chiral guest molecules impose a local twist to the longitudinal axes of the host molecules which is amplified into the bulk by the long-range molecular correlation of the solvent. Equal amounts of enantiomeric guest substances of equal optical purity induce helical structures with opposite chirality and identical pitch. This long-range chirality is responsible for the selective interaction of the cholesteric mesophases and the circularly polarized radiation. The well-known selective reflection of the two circular polarizations when the wavelength of light inside the mesophases equals the pitch of the helix is a manifestation of CIDS at 180°. This selective reflection has been investigated by numerous authors.[7-10] Nityananda and Kini[11] have obtained exact expressions for the reflection and transmission cross sections by cholesteric films in the planar orientation.

In this paper we show that a second Born approximation scattering theory can account reasonably well on a quantitative basis for the observed differential scattering cross sections in the transmitted and reflected regions of the scattering patterns.

The study of the differential scattering properties of cholesteric liquid crystals is of interest for both a formal and a practical reason. A priori, it is expected that cholesteric mesophases should exhibit rather large CIDS ratios as a function of the scattering angle. Thus, it is a unique system to confirm the predictions of the CIDS theory. Such understanding is of crucial importance in the development of CIDS into a mature optical technique. On the other hand, the use of the CIDS technique to study the cholesterics should give a new and different insight into the structure, degree of crystallinity, and order parameters of these mesophases.

We present here, therefore, results obtained using the CIDS instrument developed in our laboratories. We measured the CIDS

of both the planar and the focal conic orientations. The
patterns for the two orientations were distinct, and the
dependence of the CIDS on the handedness matched the theoretical
predictions. The details of the observed scattering patterns do
not entirely match the predictions of the simple theoretical
model used. A number of explanations for this fact are put
forward. Some of the results presented here were reported
earlier in greater detail.[12]

2. INSTRUMENTATION

The CIDS instrument used in these studies has been described
in detail.[13] Here we will only enumerate the relevant
components. These are: a) the laser, b) a Pockels cell that
produces left- and right-circularly polarized light, c) a slit
system to control the acceptance angle of the photomultiplier
detector, d) a source of modulating voltage for the Pockels cell,
e) the PMT and PMT power supply, f) lock-in amplifiers for signal
processing, and g) a chart recorder for signal display. The
photomultiplier is mounted on an arm that rotates in the
scattering plane. It is able to rotate 360°, but measurement is
possible only from 0° to 170° and from 190° to 360°. The
rotation is controlled by a goniometer that is able to move in
arbitrary intervals. For the liquid crystal measurements, the
usual interval was 2.5° to 5°. The angular resolution is
controlled by the slit system, which can be fixed by the use of
apertures and by varying the distances from the apertures to the
sample. For the liquid crystals, the angular resolution was 2°.
In these studies we used both a He-Cd laser (442 nm) and an argon
ion laser (at 488 nm) as the light sources in these experiments.
The width of the laser beam was increased by the use of a beam
expander before it was allowed to impinge upon the sample. This
is necessary to provide an area that is the effective average
over the microdomains of the liquid crystal. Using the 1.2 mm
diameter beam of the laser, the CIDS pattern was not independent
of the position of the laser on the sample. By increasing the
beam diameter to approximately 1 cm, this problem was solved.

Some instrumental difficulties were encountered in measuring
the CIDS patterns. When the polarization of the incident light
is not exactly circular (i.e., if there are linear components in
the polarization), there can be contributions to the apparent
CIDS signal from linear anisotropies in the sample. The
electro-optic modulator used to produce circularly polarized
light in our instrument creates circularly polarized light with a
stray ellipticity of about one percent. This is sufficient to
introduce some linear anisotropy contributions into our data when
the CIDS signals are smaller than about 5 percent.

3. MATERIALS AND METHODS

The right-handed cholesteric liquid crystal CB15, an
optically active p-2-methylbutyl-p-cyanobiphenyl, the left-handed
cholesteric C15, an optically active cyanoalkoxybiphenyl, and the
nematic eutectic mixture E7, a cyanoalkylbiphenyl, are
manufactured by BDH Chemicals and were purchased from E. Merck.
The nematic liquid crystal ZL1612 was purchased from E. Merck.
Polyvinyl alcohol (PVA) was purchased from Kodak. RBS 25,
(Na)p-stearyl-benzenesulfonate, was provided by Professor
Samori. All of the mixtures studied were sandwiched between a
pair of square, fused quartz plates separated by a circular 10 or
30 micron spacer (Helma Cells, Inc., 0.001 mn path length cell,
Suprasil Quartz, 210-3-QS).

The cholesteric liquid crystals were oriented with their
helix axis either parallel (planar) or perpendicular (focal
conic) to the incident beam. The orientation was determined by
the coating applied to the plates; PVA for planar and RBS 25 for
focal conic. Two sets of plates were used, one for each coating,
as the coatings were difficult to remove entirely, and using the
same plates for both resulted in poor orientation of the liquid
crystals. The plates were cleaned before each mounting by
sonication in petroleum ether (about 15 min.) to remove liquid
crystals, then soaked overnight in $Na_2Cr_2O_7/H_2SO_4$. The
plates were then soaked (about 15 min.) in dilute NH_4OH to
remove residual ions, then rinsed with deionized, filtered water
and dried with N_2 gas.

The coating of the plates with PVA was done by placing them
in a saturated filtered solution of PVA, and very slowly and
steadily pulling them out, using a variable speed stepping
motor. This dipping was done 5-6 times for each plate. The two
plates were then allowed to dry, with the side that was to
contain the liquid crystal exposed to air. After drying, they
were stroked in one direction with a piece of styrofoam to orient
the PVA. The liquid crystal was placed on one plate and the
second plate placed on top, being careful to avoid trapping air
bubbles. The second plate was oriented so that the stroked PVA
was at right angles to the stroked PVA of the first plate, to
avoid linear birefringence. The plates were placed in the cell
holder and the preparation allowed to sit for 12 to 24 hours
before measurement was made, since the orientation of the sample
was slow.

For liquid crystals in the focal conic orientation the plates
were coated with RBS 25 allowing them to stand in a 10% solution
at 65°C for about 3 minutes. The plates were then washed with
distilled water and dried with N_2. The liquid crystal was
placed between them and the cell was mounted in the cell holder.

The mounted liquid crystal was then melted with a heat gun, until it became isotropic, and allowed to cool. This was repeated two or three times, until the texture was uniform. A well-oriented focal conic alignment displayed a texture like frosted glass.

The cell holder was designed to enclose the edges of the two plates, leaving a slit in the middle approximately one cm high by 3 cm long. This puts a uniform pressure on the liquid crystal as it is pressed between the two plates, to ensure that its orientation was homogeneous. The quartz plates were mounted in the holder in such a way that no translation or rotation was possible, but that the only pressure was exerted normal to the plate surface by the washers of the holder. The holder itself was of two pieces of anodized aluminum, held together by four screws. It should be pointed out that this sample holder restricts the angles of measurement to those less than about 60° and those greater than about 120°. The measured scattering angle does not correspond to the true angle of scattering due to the refraction that occurs at the quartz-air interface. To correct for this effect we have used Snell's law with a refractive index of $n=1.5$ for quartz and 1.0 for air.

Baseline measurements were performed regularly by mounting a sample of polystyrene suspension (.109μm spheres) in the same manner as the liquid crystal sample. The baseline signal typically ranged between 0.01% to 0.1% of the full-scale signal. Most of the baseline magnitude may be attributed to stray reflections from the light entry port on the sample holder. Careful alignment of the sample cell and mounting eliminated most of these reflections, but it seems likely that the CIDS signals are still affected by stray light at angles close to 0° or 180°. A new design is currently being considered to reduce these reflection problems.

Also, linear birefringence from nonuniform PVA coatings and other defects may contribute to individual sample differences. We are currently studying techniques to control the variables that cause detectable CIDS difference from sample to sample.

Since cholesteric liquid crystals reflect light of the same handedness when the pitch matches the wavelength, it was possible, using either visible or infrared spectrophotometers, to measure the pitch of the planar samples. The sample in the cell holder was placed in the spectrophotometer with the face of the quartz plate perpendicular to the incident beam. By scanning over wavelength, the reflection spectrum of the sample was measured. The amount of reflection (or apparent absorption) and the width of the band correspond to the degree of orientation of the sample and to the distribution of pitches present.

Conversely, the absence of a pitch band for the focal conic
orientation provided evidence that there was no planar component
present. For those samples whose pitches were not measurable
(due to the IR absorbance of the quartz plates or other
experimental complications) the pitch was calculated using the
empirical relation that 5% w/w CB15 in E7 gave a pitch of 3
microns, and 20% CB15 in E7 gave a pitch of 750 nm. This inverse
relation was found to work quite well. To find the pitches of
the CB15 in ZLI 1612 mixtures we used a formula of the form

$$P = \frac{k}{(wt\%)^c}$$

We evaluated the constants k and c by first putting the
formula in a linear form,

$$\ln P = \ln k - c\ln(wt\%),$$

then plotting lnP vs ln(wt%) for four compositions whose pitches
had been measured by spectroscopic methods and finding the least-
squares slope and intercept. We obtained k and c values of 18.27
and 0.9714 respectively, with a correlation coefficient of
0.99990.

4. RESULTS AND DISCUSSION

The CB15, E7, and ZLI1612 had a specific gravity of 1
gm/cm^3, so they could be mixed by volume to give volume or
weight percentage. After addition of the nematic to the
cholesteric, the mixture was heated to its clearing point and
allowed to cool to ensure its homogeneity. C15, the left-handed
cholesteric, is a solid at room temperature. The amount of added
nematic that was necessary to keep it fluid gave a pitch of about
4 microns. Any greater C15 concentration caused the mixture to
crystallize. This difficulty made it hard to use C15 in the
planar orientation, since a 4 micron pitch is too long for stable
orientation in a 10-micron pathlength cell. Attempts to use a
30-micron pathlength cell gave no reproducible results,
presumably because 30 microns is too great a distance to maintain
a helical path that is perpendicular to the plates. The C15 was
used in the focal conic orientation, where its CIDS pattern could
be compared to that of CB15 with a pitch of 4 microns. These two
compounds are nearly identical, so a comparison of the two CIDS
patterns should not suffer from signals that might arise from
chemically different compositions. The nematic mixture E7 is
chemically similar to the cholesteric liquid crystals, which is a
critical factor in obtaining a well-oriented homogeneous sample.

When the chiral component is very different from the nematic, the two compounds will not form smooth layers, but will have discontinuities around the different shapes. This was an especially important consideration for scattering experiments, since these potential discontinuities could cause diffraction of the light.

The CIDS patterns for the focal conic orientation of C15 and CB15 are shown in Figure 1. These mixtures have similar pitches, calculated using the inverse relation of pitch to cholesteric concentration.

(helix axis perpendicular to incident light)

pitch = 4μ

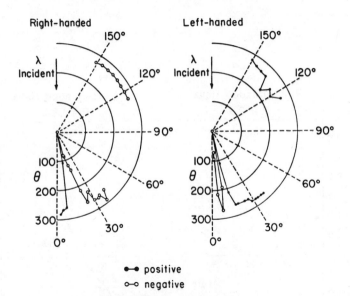

●—● positive
○—○ negative

Changing the helix handedness
reverses the sign of the CIDS.

FIGURE 1.   CIDS patterns for focal conic orientations of C15 (left-handed) and CB15 (right-handed) cholesterics, with pitch P=4μm. Notice that the differential scattering cross-sections are similar in space but of opposite sign. The wavelength of light is 442 nm. The sample pathlength is 30μm.

The lobes and zeros of the patterns are at nearly the same
angles, but the sign of the lobes is opposite. Scattering
intensities are nearly identical for the two samples, providing
some indication that the relative degree of orientation is the
same for both preparations.

Since C15 could not produce a sufficiently small pitch, a
mixture of left-handed cholesteryl-oleyl-carbonate (COC) and a
nematic (1167, from B. Samori) was used to give a pitch of 1.85
microns. The CIDS patterns of COC and CB15 in the planar
orientation are shown in Figure 2. Again it is clear that the
sign of the CIDS signal changes when the handedness of the helix
is changed.

CIDS should also be sensitive to the pitch of a helix, and
this was in fact found to be true for both planar and focal conic
textures.

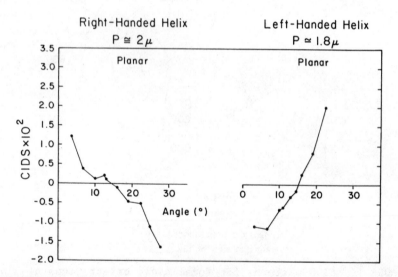

FIGURE 2. CIDS of planar cholesteric mixtures, COC in 1167
(left-handed) pitch $P=1.8\mu m$ and CB15 (right-handed) pitch
$P=2\mu m$. The wavelenth of light is 442nm.

Figure 3 shows the results for the focal conic orientation, for
mixtures of CB15 in ZLI1612 where the pitch varies from 4 microns
to 670 nm. The zeros of the scattering pattern in the forward
direction progress toward 0° as the pitch is increased. It

appears that the lobes also move as the pitch increases, but it
is not clear if they are moving forward or back. Backward
movement with decreasing pitch is likely from the data. Since,
however, there have been no computations made for an orientation
of helices analogous to the focal conic texture, such an
interpretation must be made cautiously.

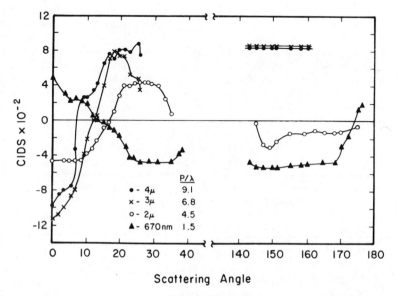

FIGURE 3. Focal conic orientation of several mixtures of
CB15 in ZLI1612. The wavelength of the light is 4.42nm.

For the focal conic texture, the magnitude of the CIDS signal
decreases as the pitch approaches the wavelength of light. The
magnitude of the CIDS signal can be influenced by a number of
factors, and it is not obvious which factor(s) have the greatest
contributions. It is clear that the degree of orientation is
critical to obtaining large signals.

The experimental data for the planar orientations of the
CB15/ZLI1612 mixture are shown in Figures 4a-4d. All data shown
in the figures were collected with an incident wavelength of
488nm, at room temperature. As can be seen, data are lacking in
the region from about 27° to 153°. Also, at angles near
0° or 180° the CIDS signals tend to decrease significantly
in all the samples. This is probably due to stray light
reflected from the sample holder.

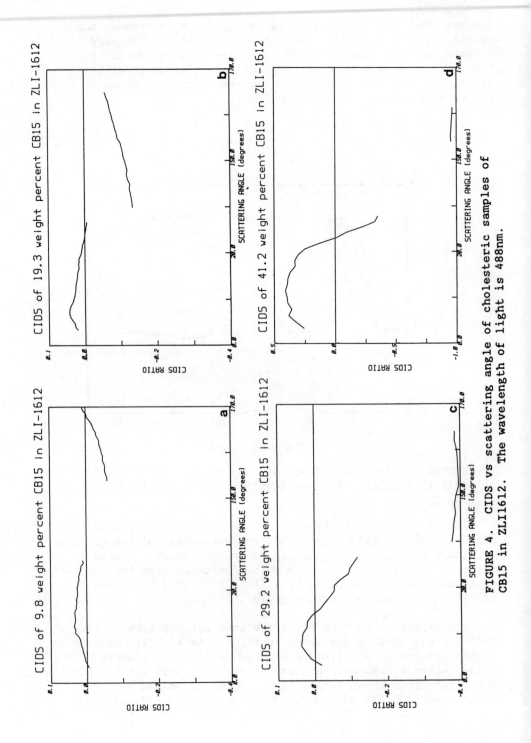

FIGURE 4. CIDS vs scattering angle of cholesteric samples of CB15 in ZLI1612. The wavelength of light is 488nm.

5. THEORY

It has been possible to obtain good agreement between our experiments for the CIDS of the liquid crystals in the planar orientation and a simple theory that makes use of the second Born approximation.[12,14,15] In our model for a planar cholesteric liquid crystal, we make use of the fact that for a periodic system the CIDS depends only on the properties of one unit cell and is not affected by the repetition of the unit cell in space. A common model for the structure of a cholesteric liquid crystal is the "stacked polarizers" model. The liquid crystal is thought of as a stack of polarizing sheets, with the polarizing axis of each sheet rotated slightly with respect to the axis of the sheet below, so that the overall stack has a helical twist along the axis perpendicular to the planes of the polarizers. The "unit cell" of this array can be visualized as a "twisted ladder", i.e., a single line of _anisotropic_ polarizable point groups with their anisotropic axes perpendicular to the lines joining them and with successive groups rotated with respect to each other. This unit cell can be represented by the polarizability density

$$\underline{\alpha}(z) = \alpha_1 \hat{t}(z)\hat{t}(z) + \alpha_2 [\underline{1} - \hat{t}(z)\hat{t}(z)]$$

where $\hat{t}(z) = [\cos \frac{2\pi z}{P}]\hat{x} + [\sin \frac{2\pi z}{P}]\hat{y}$,

and where P is the pitch of the liquid crystal, and α_1 and α_2 are polarizability density magnitudes.

It should be noted that our model is for a purely crystalline medium. In fact the liquid crystal is not crystalline, and if it were, there would be no scattering at angles other than 0° and 180°. The light we detect experimentally at angles other than 0° and 180° is therefore scattered by mechanisms not accounted for in our model, and our calculations are therefore suspect at these angles. Despite this concern, the results seem to agree reasonably well, at least at angles near 0° and 180°. A more critical test of the theory will be possible when we are able to collect data in the region between 27° and 153°.

The derivation of the CIDS for this system has been presented elsewhere,[12,15] and here we give only the results. Let the CIDS be denoted by the symbol $\Delta(\theta)$. Then

$$\Delta(\theta) = \frac{I_L - I_R}{I_L + I_R} = \frac{\Delta_o(\theta) + \Delta(0)}{1 + \Delta_o(\theta)\Delta(0)} \qquad (1)$$

where

$$\Delta_0(\theta) = \frac{x^3}{\frac{1+|\gamma|^2}{2} x^4 + \frac{1-2|\gamma|^2}{2} x^2 + \frac{1}{2} |\gamma|^2}, \tag{2}$$

and

$$\Delta(0) = \frac{-2ar^2[1 + (a-1)r^2]}{(r^2-1)^2 - ar^2[ar^2(1+r^2)-2(r^2-1)]} \tag{3}$$

with $r=P/\lambda$, $x=(\frac{r}{2})(\cos\theta-1)$, $a=(\frac{\pi}{2})(\alpha_1-\alpha_2)$, and $\gamma=(\alpha_1+\alpha_2)/(\alpha_1-\alpha_2)$. The parameters a and γ have the physical significance that they control the degree of induced coupling between scatterers in the medium and the degree of anistrophy of the scatterers, respectively. The quantity $\Delta_0(\theta)$ is the CIDS in the absence of any interactions between scatterers (i.e., in the first Born approximation) and vanishes at $\theta=0°$. As can be seen from equation 1, the quantity $\Delta(0)$ is the CIDS at $0°$.

FIGURES 5a and 5b show plots of the CIDS in the

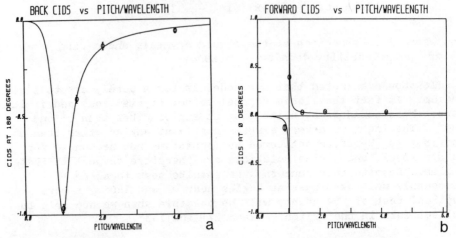

FIGURES 5a,b Backward and forward CIDS vs P/λ.
Theory: continous lines; experiments: circles.

backward and forward directions as a function of r, for both experiment and theory. The experimental values of CIDS in the "forward" and "backward" directions are actually the largest values of the CIDS in the forward and backward regions respectively. We have not used the CIDS values at precisely

5° and 170° because of the stray light problem at these
angles mentioned earlier. Theoretical calculations using eq.(1)
indicate that the CIDS should be quite constant at angles near
0° and 180°, so we expect that in the absence of stray
light the CIDS patterns at 0° and 180° would be similar
to the largest values in the forward and backward regions. For
the theoretical curves in Figures 5a and 5b the quantitites γ
and a were treated as adjustable parameters to obtain the closest
agreement with the experiments. The curves are very distinctive,
and the theory and experiment are clearly in agreement despite
some of the uncertainty of the data and the simplicity of the
theory.

6. DISCUSSION

Our purpose here has been to describe some preliminary attempts
to use circular intensity differential scattering as a method for
investigating the structure and properties of liquid crystals.
Our data are still very rough, but considerable improvement in
instrumentation and sample preparation techniques should be
possible in the near future. However, even at this stage there
are three main conclusions to be drawn from these results.
First, that CIDS can indeed indicate the sense of a helix, i.e.,
right- or left-handed. The change in the sign of the CIDS
pattern occurs for liquid crystals oriented with the helix
parallel or perpendicular to the incident beam. The theory
predicts this result, and the observation confirms the theory.

Second, the CIDS pattern is sensitive to changes in the
pitch, although the character of the response is dependent on the
alignment of the helices. For the focal conic orientation, the
front lobes appear to move backward while the back scattering
changes sign as the pitch is decreased. For the planar
orientation, the scattering dependence on pitch seems to be
related to the property of cholesteric liquid crystals of
reflecting light of the same handedness as the helix. As the
pitch approaches the wavelength of the light for the CB15 system,
the magnitude of the CIDS ratio at 180° approaches -1 as
predicted by the theory.

Third, the CIDS signals can be successfully modeled with
relatively simple theories. A major reason for the simplicity of
our expressions for the CIDS of a cholesteric liquid crystal in
the planar orientation is that in general CIDS patterns for
periodic systems depend only on the properties of one unit cell
of the periodic array.

The contributions to the CIDS pattern from inhomogeneities in
the sample arising from imperfect alignment, or variation in
pitch, are not yet clear. The main features of the patterns,

particularly for the planar orientation, are empirically
predictable, and the theory is able to predict the sign of the
back scattering. The details of the patterns, however, while
constant for a preparation over time, are more problematic to
predict by the theory.

These results give us confidence that CIDS is a sensitive
measure of chiral structure. We hope this study will not only
help us learn more about the interaction of these mesophases with
circularly polarized light, but we hope also to learn which of
these properties are shared by biological systems that display
liquid crystalline behavior.

ACKNOWLEDGEMENTS

This work was supported by National Institutes of Health
grant GM 32543 (C.B.), a 1984 Searle Scholarship (C.B.), AIO8247
(M.F.M.), GM 10840 (I.T.), and by DOE Office of Energy Research
Contract DE-AT03-82-ER60090. Partial support for this work also
was provided by a grant from Research Corporation (C.B.) and a
grant from Sandia Labs (C.B.).

REFERENCES

1. C. BUSTAMANTE, M.F. MAESTRE, I. TINOCO, JR., J. Chem.
   Phys., 73, 4273-4281 (1980).
2. C. BUSTAMANTE, M. F. MAESTRE, I. TINOCO, JR., J. Chem.
   Phys., 73, 6046-6055 (1980).
3. C. BUSTAMANTE, I. TINOCO, JR., M. F. MAESTRE, J. Chem.
   Phys., 74, 4839-4850 (1981).
4. C. BUSTAMANTE, I. TINOCO, JR., M. F. MAESTRE, J. Chem.
   Phys., 76, 3440-3446 (1982).
5. M. F. MAESTRE, C. BUSTAMANTE, T. L. HAYES, J. A. SUBIRANA,
   I. TINOCO, JR., Nature, 298, 773-774 (1982).
6. I. TINOCO, JR., C. BUSTAMANTE, M. F. MAESTRE, Structural
   Molecular Biology: Methods and Applications, eds D.B.
   Davies, W. Saenger, and S. S. Danyluk, (Plenum, New York,
   1982).
7. M.C. MAUGUIN, Bull. Soc. Franc. Miner. Cryst., 34, 71
   (1911).
8. C. W. OSEEN, Trans. Faraday Soc., 29, 883 (1933).
9. H. DE VRIES, Actn. Cryst., 4, 219 (1951).
10. E. I. KATS, Sov. Phys., JETP, 32, 1004 (1971).
11. R. NITYANANDA, V. D. KINI, Proceedings of the
    International Liquid Crystal Conference, Bangalore,
    India, (December, 1973) (Pramana Supplement I,).
12. C. BUSTAMANTE, K. S. WELLS, D. KELLER, B. SAMORI, M. F.
    MAESTRE, I. TINOCO, Jr., Molecular Crystals and Liquid
    Crystals, (in press, 1984).

13. J. KATZ, K. S. WELLS, D. USSERY, M. F. MAESTRE, C.
    BUSTAMANTE J. of Sci. Inst, (accepted, 1984).
14. C. BUSTAMANTE, M. F. MAESTRE, D. KELLER, I. TINOCO, JR.,
    J. Chem. Phys., 80, 4817-4823 (1984).
15. D. KELLER, Ph.D. Thesis, Department of Chemistry,
    University of California, Berkeley (1983).

CIRCULAR INTENSITY DIFFERENTIAL SCATTERING OF NATIVE

CHROMATIN: THEORY AND EXPERIMENTS

Claudio Nicolini

Chair of Biophysics

Faculty of Medicine University of Genova Italy

Due to their particulate nature, native biopolymers (such as membrane proteins or chromatin) tend to form an intensely light-scattering suspension. Scattered light is deviated from the beam axis at an angle great enough to miss the detector, resulting in (a) anomalously high optical density and (b) circular intensity differential scattering (CIDS), i.e., left and right circurlarly polarized light are scattered with different efficiences out of the photomultiplier (PM) window.

It must be enphasized that strict unpolarized light-scattering (absorbance measurement, OD) and polarized light-scattering (CD) outside the absorption band are two different phenomena and the latter appears to have greater structural significance in the study of macromolecular structures (such as membrane, ribosome or chromatin), as discussed in this workshop.

Found originally in native chromatin (1,2) and intact T2 bacteriophage (3), and more recently in intact mammalian nuclei, the CD signal outside the absorptive band (305 nm) appears highly dependent from PMT-sample distance and disappears with shearing - which is known to disrupt quaternary structure (2) - even despite the persistence of a relatively high strict light scattering (Fig.1). The relation between observed ellipticity and wavelenghth (above 305 nm) was originally investigated by empirically fitting the observed data to a simple exponential model (1). The

results indicated that the observed long wavelenghth
(300 nm) data were satisfactorily fitted by a scat = A +
B (correlation coefficients greater than 0.95).

The values of the costant A (possibly due to "inactive
sample" artifacts attributable to instrumental misalignment)
and the slope B (which contain structural information and
are highly dependent on cell type), obtained by
least-square fits to the observed ellipticities outside the
absorptive band, were used not only to compute the true CD
in the absorptive band (correcting for such artifacts), but
also were used to probe (the slope B namely) for the
quaternary chromatin-DNA structure (2,5).

Recent theoretical analyses on the nature of this
phenomena in chromatin solutions and on its dependence from
quaternary chromatin structure recently introduced (6,7)
have indeed given solid foundations (from first physical
principles) to the original empirical observations and to
these early gross explanations ,promising further exciting
developments.

EXPERIMENTAL DATA

The phenomenon of differential scattering of left
versus right handed circularly polarized light (CIDS) has
been observed in isolated "native" chromatin, as a spurius
positive ellipticity outside the absorption bands (above
300 nm) of either protein or DNA in three biologically
significant experimental situations (1,2,5). First , under
conditons of equivalent total light scattering, as measured
by the OD (360)/OD(260) ratio, where there existed a strong
CIDS effect for unsheared chromatin, the CIDS for the same
but sheared chromatin remained zero. Second, in unsheared
chromatin, the CIDS was found to be significantly lower for
"S" phase than for G1 phase HeLa cells, and to be reduced
for fibroblasts three hours after serum stimulation to 1/20
of its control value (G0 versus G1 phase cells). Third,
using the light microscope, measurements of CIDS have been
made in situ from individual nuclei (7a). Consistent with

the earlier findings, a large CIDS was found in Gl cells with small dense nuclei while it was virtually absent in early S cells with large and dispersed nuclei.

These observations highlighted the potentail for CIDS as a possible probe for studying the higher order structure of chromatin and its modification in the control of cell function and gene expression (8).

However, a realization of this potential required a satisfactory theoretical formulation for the phenomenon of CIDS, as applied to chromatin or membranes which was lacking until recently.

Recently, Bastamante, Maestre, and Tinoco (9) have developed analytically an expression describing CIDS from a helix with uniaxal polarizability directed tangent to the helix in the Born approximation (equivalent to a dipole approximation). While clearly demonstrating the possibility of CIDS arising solely from the simmetry properties of the overall structure of the scattering center, its direct relevace to chromatin is questionable. A true continuous helix is applicable only to the DNA within chromatin. Moreover, the DNA superhelix is present in the individual nucleosome, and it is therefore not clear why the CIDS phenomenon should be absent after shearing or in disperse nuclei if its origin was solely from the DNA (in both cases the tertiary nucleosome structure - with a DNA supercoiled aroud the octamer histones - is indeed preserved).

An even more relevant argument is that of the difference of size of the protein nucleosome core compared to that of the DNA component. It has been previously (1) demonstrated a dependence of the CIDS phenomenon. This implied that its origin was in the dipole component of the scattered fields. Moreover, dipole radiation both electric and magnetic, varies roughly as a , where a is the characteristic radius of the scattering center. It therefore appeared reasonable that the dominant contribution of the CIDS was from the nucleosome core. It is within this assumption that we have recently investigated this possibility by computing the CIDS from a chiral arrangement of nucleosomes ,starting from the first physical principles of classical electrodynamics.

THEORETICAL SIMULATONS

Two theoretical approximations were undertaken (6, 7). The first is to drop the Born approximation which assumes the difference in dielectric constants between the scattering center and the surrounding medium to be small. Essentially, this approximation is equivalent to treating the entire scattering object as an arrangement of independent dipole scattering centers, where each center is small relative to the wavelength, and the fields at each scattering center are assumed equal to that of the incident light.

In the 300 A° chromatin fiber, where the nucleosomes are closely opposed to each other, the validity of this approximation is not clear.

Furthermore since dipole radiation (both electric and magnetic) varies roughly as a , where a is the characteristic radius of the scattering center, the Born approximation is usually considered a good approximation only if the spacing of the scatters is at least three times the characteristic radius of each. In the various models of chromatin now being proposed this is not the case. For instance, in a typical model of the superhelix, the nuclesome spacing is approximately 55 A° (6 Nucleosomes/term, 100 A° helical radius, 110 A° pitch) between centers, while the equivalent spherical radius of the nucleosome (having the same volume as the particle) is 65 A°. Thus the close packing of the nucleosomes (scatterers) indicates that the independence of the dipole scatters can be assumed valid only to a first approximation.

Multiple Scattering of Dipoles

Once the Born approximation is dropped, most geometric arrangements of scattering centers are analytically intractable. Therefore, one must resort to computer simulations. Even here, if one was to simulate the boundary value problem, it would be computationally intractable.

Instead, in this first approximation (6) we utilize an approach first proposed by Purcell and Pennypacker (10) for studying the light scattering propoerties of irregularly shaped interstellar dust. In that work they approximated an irregular shaped grain by a lattice of spherical dipole oscillator which are assumed to interact. That is, retardation effects are taken into account by including in addition to the incident radiation, the contribution of the j-th oscillating dipole to the electric field at the site of the i-th dipole. A self-consistent set of (complex) electric fields at the i-th dipole is thus sought.

We utilize this approach in the following way. We assume that we have a helical array of spherical nucleosomes, each nucleosome corresponding to a dipole scattering center. We then solve the resulting linear equations for the self consistent electric fields at each nucleosome assuming either left or right circularly polarized incident light. After obtaining this, we then calculate the scattered light at various points in space representing detection positions and obtain the amount of differential light scatter after averaging for the possible orientations of the molecules (6).

This allows us to simulate the experimental data, namely the absence of CIDS for sheared "nucleofilement-like" chromatin and its presence for unsheared "solenoid-like" (11) or "rope-like" (12) native chromatin fibres. The model actually does predict CIDS for a helical arrangement of multiple dipole scatters for any given orientation of the helix with respect to the incoming light. The effect of averaging over all spacial orientations of the helix (which mimics the experimental situation) is shown in Figure 1 ,where the expected symmetry (lacking for any given orientation) in the angle dependence of CIDS is found.

Similarly the model can explain the structural alterations between late G1 and middle S-phase which cause significant but finite differences in differential light scattering in isolated bulk chromatin which then becomes a yes (G1 nuclei) versus no (S nuclei) phenomenum in individual nuclei. The sign (Figure 1) and magnitude of the

differential scattering of left versus right circularly
polarized light (CIDS), outside the DNA absorptive band, is
explainable only with a "left-handed" helical folding of one
or more nucleofilaments (with the "rope" giving a higher
(CDLS) signal than the "solenoid"); furthermore transition
from a highly supercoiled nucleofilament to a linear
"unfolded" nucleofilament would be accompanied by a decrease
in CIDS down to a null differential light scattering, as
indeed experimentally found for S phase nuclei.

Few of the most exciting results of the simulations are
the relative insensitivity of the the CIDS signal to
variations of the length of the biopolymer and its high
sensitivity to any variation in the pitch of the biopolymer
superhelix.

Born approximation

In a second approach (7) we have added complexity by
allowing the nucleosome a more realistic shape, but assuming
independence among the scattering centers (within the Born
approximation). In this respect, we were aided by nature
having given the nucleosome a near oblate ellipsoid shape,
as the dielectric ellipsoid is one of the few analytically
tractable shapes other than the sphere to which the dipole
approximation can be readily applied. By using the ellipsoid
shape, the individual nucleosomes no longer represent
identical scatterers with respect to the incident light
direction (except for particular orientations of the
nucleosomes in relation to the helical axis in the
superhelical nucleofilament folding) and the possibility
exists for CIDS, even within the Born approximation (7).

For the calculations presented in this case (7) , we
assume that the particles are shaped from an isotropic,
dielectric material with a scalar dielectric constant. In
this case, it can be shown that as a result of the
ellipsoidal shape the polarizability of the particle is a
tensor, whereby the induced particle dipole moment can be
expressed as a tensor equation in terms of an appropriate
orthogonal coordinate system - chosen with axes oriented
parallel to the axes of the ellipsoid (7) .

In this tensor notation the electric field scattered
from the j ellipsoid can now be written (in the far

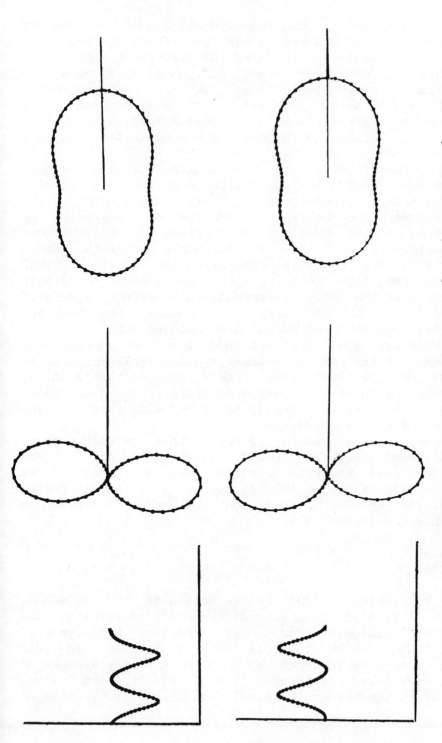

FIGURE 1. Model simulation for a helix 110 A° pitch and 100 A° radius
left-handed (above) and right-handed (below). (Left) CIDS versus angle;
(Center) Polar plot of the absolute light scatter versus angle: (Right)
Polar plot of total light scatter versus angle. From references 6
with permission.

field approximation) as the vector sum of two linearly independent fields, and the intensity of the scattered light is finally calculated by adding the electric fields at the observation point from each ellipsoid. The resulting differential scattering cross section -theoretically computed - closely parallel the one experementally determined under similar configuration (7,1,2).

All calculations were carried out numerically using a digital computer.

As a general feature of this simulation within the Born approximation, the numerical values of the CIDS signal - under similar geometrical configuration of the polynucleosome superhelix - confirm qualitatively the conclusions of the previous simulations "outside the Born approximation" , but are significantly larger than those derived from the multiple scattering of dipoles . This points to the fact that the interdependence of the dipole scatterers may be only a second-order effect - even if important and to be taken into consideration in any theoretical simulation of actual CIDS measurements.

Furthermore the CIDS obtained for any superhelical arrangement of the DNA alone appears quite small compared to the CIDS of the nucleosome itself ,showing that a chiral arrangement of non-chiral but anisotropic subunits is needed to produce the experimentally observed differential light scattering of native chromatin

It may be also worthy to notice that the CIDS signals of significant magnitude could be obtained only if the nucleosomes were oriented parallel to the tangent to the helmix ,a situation strikingly consistent with most recent experimental work (for a complete updated review on chromatin see reference 5).

CONCLUSION

In conclusion , due to the origin of the phenomenon, the CIDS (circular intensity differential scattering) for the single largest and most important biopolymer (chromatin-DNA ,which controls gene expression and cell function) appears to depend more on the simmetry aspects of the macromolecular structure (i.e., the pitch and helical radius of the quaternary folding) than on the acutal size of the macromolecule, instead critically dependent on the total light scattered.

In this way, information related to the higher order structural parameters might be obtained even from a heterogeneous length distribution of chromatin fibers - or any other macromolecular structure for that matter - allowing interferences concerning in situ structure, as well as measurements on small quantities of isolated macromolecule (without the need for prior fractionation according to size, typical of most other structural probes).

Acknowledgement

This work was supported by a grant from the National Research Council, Finalized project on "Oncology" .

REFERENCES

1) Nicolini, C. and Kendall, F., PHYSIOLOGICAL CHEMISTRY AND PHYSICS ,9 ,265 (1977)

2) Nicolini, C. ,Kendalla,F. and Baserga,R. ,SCIENCE ,192 ,796 (1976)

3) Dorman ,B. and Maestre,M. ,P.N.A.S.-U.S.A. ,70 ,251 (1973)

4) this volume

5) Nicolini, C. ,ANTICANCER RESEARCH ,3 ,63 (1983)

6) Zietz,S. ,Belmont ,A. and Nicolini, C. , CELL BIOPHYSICS ,5 ,163 (1983)

7) Belmont ,A. ,Zietz ,S. and Nicolini, C. ,BIOPOLYMERS , in press

8) Nicolini, C. ,"BIOPHYSICS AND CANCER" , Plenum Publishing Co.,New York (1984)

9) Bustamante,C. , Maestre,M. and Tinoco ,I., J. CHEMICAL PHYSICS ,73 ,4273 (1980)

10)Purcell,P. and Pennypecker ,A. ,ASTROHYSICS J. ,186 ,705 (1973)

11)Finch,J. and Klug, A., P.N.A.S. -U.S.A. ,73 ,1897 (1976)

12)Nicolini,C. ,Beltrame,F. ,Cavazza ,B. ,Trefiletti,V. and Patrone ,E.,J. CELL SCIENCE ,   (1983) 62,103

MULTIPARAMETER LIGHT SCATTERING

FOR RAPID VIRUS IDENTIFICATION

Gary C. Salzman, W. Kevin Grace,
Dorothy M. McGregor and Charles T. Gregg

Experimental Pathology Group
Life Sciences Division
Los Alamos National Laboratory
Los Alamos, NM 87545

INTRODUCTION

Multiparameter light scattering (MLS) is a term we have used to describe the simultaneous measurement of multiple elements of the Mueller matrix at specific wavelengths and scattering angles. This 4 X 4 matrix describes the polarization sensitive transformation of an incident beam of light into a scattered beam of light by a scattering object such as a virus or suspension of virus particles. The Mueller matrix contains a great deal of information about the internal structure and shape of the virus particle. This information is sufficient in many cases to enable discrimination among a wide variety of different viruses of clinical significance.

The purpose of this paper is to describe a photopolarimeter for making simultaneous angular distribution measurements of up to eight of the Mueller matrix elements from a suspension of virus particles.

PHOTOPOLARIMETER

Fig. 1 shows a schematic of the photopolarimeter. The 488 nm light from the argon laser (50 - 250 mW) passes through a glan-laser prism polarizer (POL1) to ensure that a high fraction of the transmitted light is linearly polarized at a specific angle with respect to the horizontal scattering plane. The beam then passes through a photoelastic modulator [MOD(F1)] operating at 50 kHz (Hinds International, Portland, Oregon). The beam incident on the sample is elliptically polarized with its handedness alternating left and right at 50 kHz. The incident intensity is not modulated. The sample is contained in a cylindrical quartz cuvette 50 mm in diameter with 2 mm thick walls. The entrance window is a 0.17 mm thick piece of coverslip. A Rayleigh horn is used to dump the transmitted beam. The cuvette is cut from a piece of extruded quartz tubing so its optical properties are not optimal.

103

MULTIPARAMETER
LIGHT
SCATTERING
(MLS)

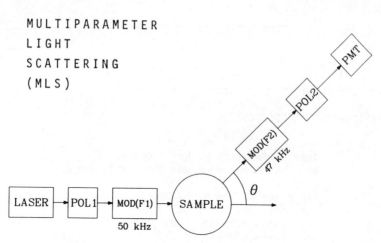

Fig. 1.   Schematic   of   the   multiparameter   light   scattering
photopolarimeter.

    The telescope to detect the scattered light is mounted on a
computer controlled rotary stage.   The acceptance half angle of the
telescope is 0.5 degrees.   The scattered light passes through a second
photoelastic modulator [MOD(F2)] operating at 47 kHz. Modulation at a
second frequency makes it possible to measure separately each of the
Mueller matrix elements rather than linear combinations of the elements
as some other investigators have done.   The scattered light then pases
through a second linear polarizer (POL2), a laser line filter and
impinges on the photomultiplier tube (PMT).   This photopolarimeter is
conceptually similar to one developed by Thompson, et al [1] that uses
four electrooptic modulators and can measure simultaneoulsy all 16
elements of the Mueller matrix from a suspension sample.   This
instrument is in turn based on a single modulator instrument developed
by Hunt and Huffman [2] and used by Bickel, et al [3] to make measurements
on a number of bacterial and viral suspensions.

    Fig. 2A shows the coordinate system used in the photopolarimeter.
The azimuthal angle convention follows that used by Bohren and Huffman
[4]. Fig. 2B shows a schmatic top view of the photopolarimeter with the
azimuthal angles indicated for each of the optical elements.   The
retardance, δ_i, of each modulator is set at 2.4048 radians.   At this
retardance the total scattered intensity, which is used to normalize
the other Mueller matrix elements, is independent of the polarization
of the scattered light.

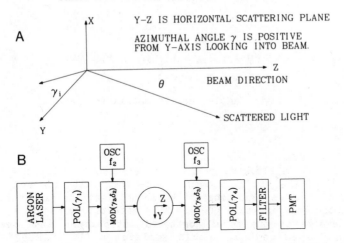

Fig. 2. (A) Coordinate system used to describe the MLS photo-polarimeter. (B) Schematic top view of the photopolarimeter showing the parameters for each of the optical elements.

With two modulators eight of the matrix elements can be measured simultaneously. The two modulation frequencies produce intensity modulation at the PMT at a variety of frequencies consisting of linear combinations of the two input frequencies. Each element of the Mueller matrix appears at a unique modulation frequency. With the azimuthal angles of the passing axes of POL1 and POL2 at +45° and the azimuthal angles of the fast axes of MOD1 and MOD2 at +90° (case I) the measurable Mueller matrix elements are as shown in Fig. 3. The numbers in parentheses are the frequencies in kilohertz at which each matrix element is measured. The dashes in some of the parentheses indicate that there is no contribution from that particular matrix element.

MUELLER MATRIX

S_{11}(DC) S_{12}(-) S_{13}(100) S_{14}(50)

S_{21}(-) S_{22}(-) S_{23}(-) S_{24}(-)

S_{31}(94) S_{32}(-) S_{33}(6) S_{34}(44)

S_{41}(47) S_{42}(-) S_{43}(53) S_{44}(3)

Fig. 3. Sample Mueller matrix for case I.

With the polarizer passing axis azimuthal angles at +90° and the modulator fast axes at +45° (case II), the Mueller matrix is as shown in Fig. 4. Note that row two and column two are not present with this configuration.

MUELLER MATRIX

$S_{11}(\ DC)\quad S_{12}(100)\quad S_{13}(\ -\ )\quad S_{14}(\ 50)$

$S_{21}(\ 94)\quad S_{22}(\ 6)\quad S_{23}(\ -\ )\quad S_{24}(\ 44)$

$S_{31}(\ -\ )\quad S_{32}(\ -\ )\quad S_{33}(\ -\ )\quad S_{34}(\ -\ )$

$S_{41}(\ 47)\quad S_{42}(\ 53)\quad S_{43}(\ -\ )\quad S_{44}(\ 3)$

Fig. 4.   Sample Mueller matrix for case II.

Fig. 5 shows a block diagram of the electronic circuitry needed to measure two of the matrix elements.   As the scattering angle changes the anode current is kept constant by controlling the high voltage bias on the photomultiplier tube (PMT).   This effectively normalizes all the matrix elements to the total scattered intensity.   The preamplified signal is fed to several lockin amplifiers (EGG-PAR 5301).   The PDP-11/44 computer controls the rotary stage (Aerotech) and the lockin amplifiers over an IEEE-488 bus.   The 44 kHz frequency needed for the S_{34} matrix element is generated from the two input oscillator frequencies with the mulitpliers and filters shown in Fig. 5.

CALIBRATION

Each of the Mueller matrix elements is calibrated by rotating the detector arm to 0° and replacing the sample cuvette with various combinations of optical elements whose Mueller matrices are known.   For

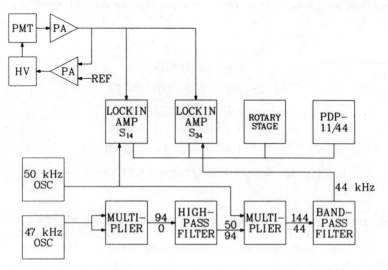

Fig. 5.   Block diagram of the electronic circuitry used for measuring the S_{14} and S_{34} Mueller matrix elements.

example S_{14} and S_{34} are calibrated by inserting a 488 nm quarter wave plate with its fast axis at 0° followed by a polarizer. The polarizer is rotated in steps through azimuthal angles ranging from 0° to 180°. S_{14} behaves as the sine of twice the azumuthal angle and S_{34} behaves as the sine squared of twice the azimuthal angle. These calibrations are shown in Figs. 6 and 7. These calibrations also enable the determination of the correct phase offsets between the reference and signal waveforms.

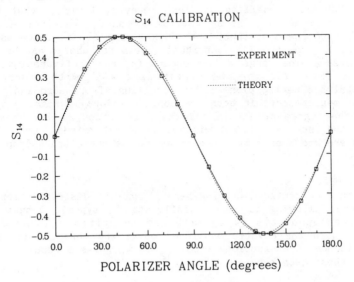

Fig. 6. S_{14} calibration

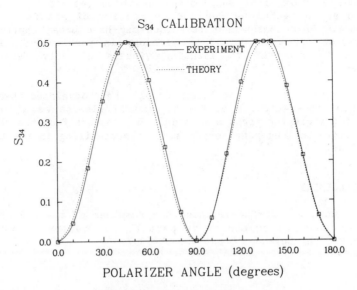

Fig. 7. S_{34} calibration

CUVETTE BIREFRINGENCE

Ideally the cuvette should exhibit no strain birefringence at either the input or output windows. A strain of less than 10 nm of retardance at a specified wavelength per centimeter of windown material is condidered strain free in the optics industry. S_{14} is affected only by strain in the input window while S_{24} is affected equally by strain in the input and output windows. Strain in the input window causes S_{12} and S_{13} to have contributions at the same frequency as that for S_{14}. Since S_{12} can be as large a -1 (at 90°) this presents a potentially severe problem. Modelling studies, however, indicate that for input window strains as large as 24.4 nm/cm and an S_{12} value of -0.5 and S_{14} value of 0.001the contribution at 50 kHz from S_{14} is twice as large as that from S_{12} and the S_{12} contribution does not change as the S_{14} value varies over a wide range. This occurs for the configuration in Fig. 3 when the fast axis for the strain is a $+45^{\circ}$ with respect to the horizontal scattering plane. Our modelling effort also indicates that a set of measurements in both the configurations in Figs. 3 and 4 will enable the degree of strain and the orientation of its fast axis in each of the input and output windows to be determined. Once these four numbers are known each matrix element can be corrected appropriately.

RESULTS

We have presented data elsewhere[5,6] showing discrimination with S_{14} among the following viruses: influenza (3 types), dengue fever (4 types), encephalitis (4 types), and hepatitis. These data were obtained with the cuvette mentioned above and were not corrected for strain induced contributions from S_{12} or S_{13}. As a result we will not present these data here.

CONCLUSIONS

An instrument for making multiparameter light scattering (MLS) measurements from viral suspensions has been described. Calibration methods and a technique for the correction of artifacts was also presented. This measurement technique may have broad applications in cell biology in general and in virology in particular. It may be possible to use MLS to identify viruses directly from clinical specimens.

Bustamante, et al [7] and Zeitz, et al [8] have developed theories that may explain the behavior of the S_{14} matrix element as a function of scattering angle for virus specimens. S_{14} appears from the theories to be sensitive to long range order such as supecoiling in the DNA or RNA in a virus.

ACKNOWLEDGEMENTS

This work was performed under the auspices of the U.S. Department of Energy and was supported in part by the National Institute of General Medical Sciences Grant No. GM26857.

REFERENCES

1. R. C. Thompson, J. R. Bottiger, and E. S. Fry. Measurement of polarized light interactions via the Mueller matrix. Appl. Opt. 19, 1323-1333 (1980).

2. A. J. Hunt and D. R. Huffman. A new polarization-modulated light scattering instrument. Rev. Sci. Instrum. 44, 1753-1762 (1973).

3. W. S. Bickel, J. F. Davidson, D. R. Huffman, and R. Kilkson. Application of polarization effects in light scattering: a new biophysical tool. Proc. Nat. Acad. Sci. USA 73, 486-490 (1976).

4. C. F. Bohren and D. R. Huffman. Absorption and scattering of light by small particles. John Wiley and Sons. (1983).

5. G. C. Salzman, W. K. Grace, D. M. McGregor and C. T. Gregg, An instrument for virus identification by polarized light scattering: a preliminary report. In: Proceedings of the 4th International Symposium on Rapid Methods and Automation in Microbiology. June 7-10, 1984, Berlin. Springer-Verlag (in press).

6. C. T. Gregg, D. M. McGregor, W. K. Grace and G. C. Salzman, Rapid Identification by circular intensity differential scattering. ibid.

7. C. Bustamante, I. Tinoco, Jr., and M. F. Maestre. Circular intensity differential scattering of light. IV. Randomly oriented species. J. Chem. Phys. 76, 3440-3446 (1982).

8. S. Zeitz, A. Belmont and C. Nicolini. Differential scattering of circularly polarized light as a unique probe of polynucleosome superstructure. A simulation by multiple scattering of dipoles. Cell Biophys. 5, 163-187 (1983).

HIGH SPEED PHOTOELASTIC MODULATION

William H. Rahe, Robert J. Fraatz and Fritz S. Allen

Department of Chemistry
University of New Mexico
Albuquerque, NM 87131

ABSTRACT

An inexpensive and easily constructed device for the high
speed rotation of the plane of polarization of a beam of
monochromatic plane polarized light has been demonstrated.The
device can be successfully used as an optical chopping device
for polarized light at speeds between 20 KHz and 4 MHz or to
introduce any other retardance leading to circular or elliptical
polarization forms.   The   optical   extinction/transmittance
pattern for the half-wave operation, although not truly
sinusoidal, has been mathematically determined. To drive the
modulation, a high precision sinewave generator capable of
automatic gain and frequency compensation is recommended;
however, the device can be used successfully with almost any
sinewave generator. Voltage requirements increase as the
wavelength of the light increases and varies between 10V and
200V dependent upon the wavelength of light used and the
multiple of the fundamental drive frequency desired. These
voltages can be obtained from many wideband amplifiers currently
available.

INTRODUCTION

Many spectroscopic experiments utilize optical chopping
methods in con junction with phase-lock detection electronics to
enhance the signal to noise (S/N) ratio of the data obtained.
Mechanical choppers are limited by their physical nature to
chopping speeds below ~ 3 KHz. Many experiments involve
measurements of optical events which occur too rapidly to take

advantage of the mechanical chopping devices currently available. Transient dichroism measurements, as an example, occur in time periods as short as 50 μsec. A higher speed alternative for optical chopping utilizing polarization modulation techniques is available for systems which can employ plane polarized light.

A diagram of such an optical system is shown in Figure 1. Clearly, differing relative alignment of two polarizers in a monochromatic beam results in differing transmitted light intensities. Regular rotational exchange of the plane of polarization from that established by the final polarizer by 90° relative to the axes of the polarizers results in alternating conditions of maximum transmission and total extinction. For optical systems utilizing polarized beams, this effect can be used in a similar fashion to conventional mechanical choppers. The Pockels cell is a device capable of performing this type of rotation.

The Pockels effect is due to the fact that several uniaxial crystals (e.g., $NH_4H_2PO_4$ and KH_2PO_4) show induced linear birefringence whose magnitude is a function of the strength of an external electric field. By attaching thin metallic grids or semi-transparent electrodes to the crystal, such that the electric field and the optical axis of the crystal are parallel to the propagation direction of the light beam, and by placing the device between two crossed polarizers, the intensity is observed to increase as an external electric field is applied to the two electrodes. This effect has been used with excellent results as an extremely fast switch for lasers. Pockels cells have been used to modulate light at a less than nanosecond time scale.[1] A representative voltage needed to rotate the plane of polarization by 90° is 5 kV. High frequency (> 20 KHz) sine waves at these voltages are difficult to generate and couple with the Pockels cell. Consequently, an alternative method has been developed utilizing strain birefringence.

A photoelastic modulator is an electro-optic device in which the optically transparent medium is an amorphous material subject to induced strain birefringence, an effect well known to glassblowers. A non-birefringent material such as fused silica can be distorted by strain from a piezoelectric drive unit into a uniaxial formation which is birefringent. By properly grinding the optical element to a length resonant with the piezoelectric driver, one minimizes the power requirements necessary to rotate the plane of polarization by 90° in either direction. The mathematics of electro-optical devices including acousto-optic and photoelastic modulators have been discussed by Kemp[2] and Kaminow.[3]

Commercial photoelastic modulators are presently available with resonant frequencies between 20 KHz and 80 KHz.[4] The voltage required for $\lambda/2$ retardation depends linearly upon the wavelength of light and is typically in the 10–50 V range. Since one vibrational period of the optical element encompasses a stretch and compression strain, two optical extinctions or transmission segments are observed for a single vibrational period. Hence, the optical modulation frequency is twice that of the crystal drive frequency.

For some experiments, the desired modulation frequency is greater than the frequencies obtainable from commercial devices. In the electric dichroism experiment used in this laboratory modulation frequencies of 500 kHz or greaterare desired. It should be pointed out that the fundamental frequency of the crystals and the physical size of the optical window vary inversely. Thus, the upper limit to the speed of modulation is restricted by the size of the light beam. In this application, the optical beam of light is 1 cm in height and 2 mmin width and increasing the modulation frequency by crystal selection is severely limited.

CONSTRUCTION OF A PHOTOELASTIC MODULATOR

A piezoelectric crystal with a fundamental frequency of 125 KHz was ordered[5] such that the maximum acoustic transfer path is in the longitudinal axis (–18.5° x-cut). The dimensions of this crystal were approximately 1 cm x 1 cm x 2 cm. Because the speed of sound is somewhat faster in quartz than in fused silica, an optical block of fused silica was ordered with dimension 1 cm x 1 cm x 2.5 cm. The fused silica block was bonded to the piezo-electric crystal using a very thin layer of epoxy. Since the resonant frequency of the optical unit is a function of its length, it was necessary to grind the optical unit until the resonant frequency of both halves was identical. Optimization was performed by minimization of the drive voltage while observing the frequency shift as the optical component was shortened by grinding the long axis with carborundum.

The sine wave necessary to drive the photoelastic modulator was obtained with a Harris PRD 7828 frequency synthesizer coupled to an external amplifier to obtain the voltages required. A tube-type audio amplifier (Bogen Model MO100) was modified to allow for a wider bandwidth by removal of the output audio transformers. Substantial improvements were made to the power supply of the Bogen amplifier to reduce the 60 Hz and 120 Hz noise. Using this amplifier, voltages of 150 V_{peak} to peak at a frequency of 125 KHz were obtained.

EXPERIMENTAL CHARACTERIZATION

After the frequency doubling previously described, the optical chopping frequency for λ/2 retardation is still only 250 kHz and as such does not meet the requirement of 500 kHz. However, the degree of retardation is a monotonic function of the voltage applied to the piezoelectric crystal, i.e., the drive voltage is doubled to change the optical retardation from λ/2 to λ/1. Subsequent voltage increases result in 3λ/2, 2λ, etc. With each of these steps, a multiplication of the fundamental frequency is observed in the optical signal.

As the frequency multiple increases, it is apparent from the observed optical signals that the period between extinctions is not linear in time. This is due to the fact that the voltage required to generate the stress necessary to cause a nλ/2 retardation is a function of the amplitude of the sine wave at any time in its period. If sufficient voltage is supplied to cause a 3λ/2 rotation, the time intervals between the voltages required for λ/2, 2λ/2, and 3λ/2 extinctions are not equally spaced in time.

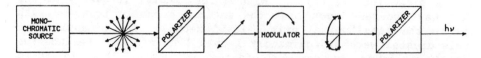

FIGURE 1: Optical Modulation System

The high speed modulator was placed in the optical train (Figure 1) between two crossed polarizers so that no light passed in the absence of rotation by the modulator. Applied voltages for ± nλ/2 (1 ≤ n ≤4) retardation were determined in intervals over the region 220 nm to 400 nm. This region was chosen since this was the region of interest for the application and the results obtained should remain valid through the visible and into the infrared regions. Results are shown in Figure 2. During this test, the fundamental frequency was observed to drift slightly when higher voltages were applied. This phenomenon is attributable to thermal changes in the crystals. Since this drift is relatively slow, it should create no difficulties in data reduction and the use of feedback loops in the drive circuitry should provide adequate stabilization. This technique is used in the commercial modulators but was not included in the drive circuits for the testing of the high speed modulator.

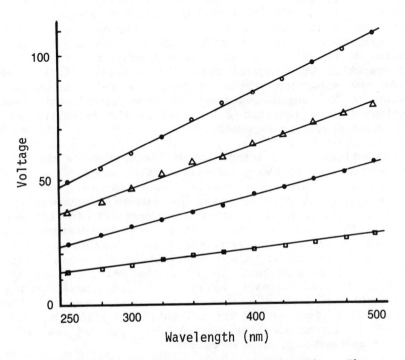

FIGURE 2: Modulation Voltage versus Wavelength. The squares
          are for λ/2 modulation, the closed circles for
          2λ/2   modulation,   the   triangle   for   3λ/2
          modulation,   and   the   open   circles   for   4λ/2
          modulation.

As indicated earlier, non-sinusoidal behavior is observed as the optical modulation frequency increases. We have described this modulation as follows:

$$\theta = A \pi \sin (\omega t)/2 \qquad (1)$$
$$I = \cos^2\theta. \qquad (2)$$

In equation 1, the angle θ represents the hypothetical angle of rotation of the plane of polarized light produced by the modulator, where A is the power parameter from the drive circuitry, ω is the fundamental circular frequency of the crystal and t is time. θ is a real angle of rotation for the $n\lambda/2$ operations described here. When A has unit magnitude, equation 1 describes the half wave operation. For A=2, full wave operation is obtained, etc. The intensity, I, transmitted by the two polarizer system is described by a cosine squared function. The angle generated by equation 1 can then be substituted into equation 2 to produce the intensity vs time curve which should be expected.

Theoretical versus actual modulation traces are shown in Figures 3-4 for each of the $\pm n\lambda/2$ retardation states ($1 \leq n \leq 4$). The overlap between the modeled function and the observed optical output is excellent and is estremely encouraging with regard to the use of phase-lock processing in the digital domain. We have developed this post-measurement digital technique to emulate the performance of a lock-in amplifier for our rapid transient signals from electric dichroism.[6] For this purpose, it is important to know the form of the carrier signal. An additional method for post-measurement S/N improvement has been described in a companion paper.[7] The use of the modeled form of the optical intensity signal has led to a new method of calibrating photoelastic modulations which is given in another part of this volume.

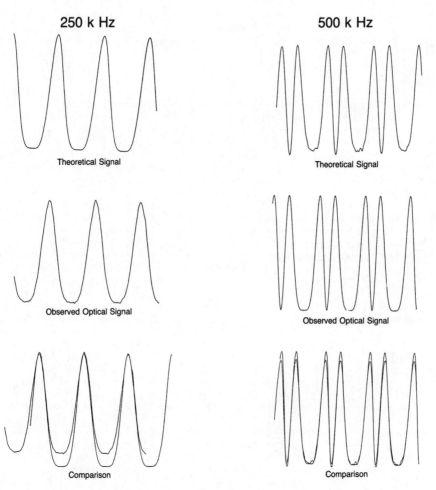

FIGURE 3: A Comparison of Measured and Calculated Modulated Signals

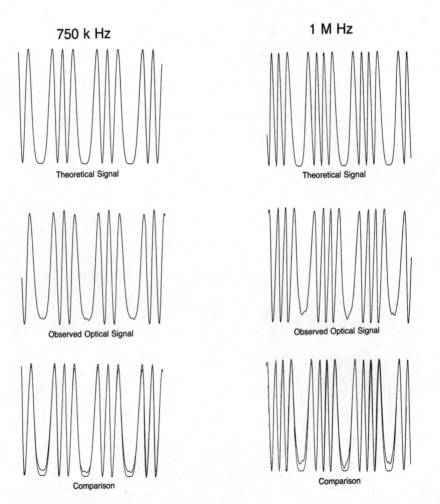

FIGURE 4: A Comparison of Measured and Calculated Modulated
           Signals

ACKNOWLEDGEMENTS:

We are grateful for support from the National Institutes of Health (grant #GM 27046) and from the Research Allocation Committee of the University of New Mexico.

REFERENCES

1. Jenkins, F. A. and White, H. C. Fundamentals of Optics, McGraw-Hill, NY pp. 695 (1976).

2. Kemp, J. C. J. Opt. Soc. Am., 59, 950 (1969).

3. Kaminnow, I. P. An Introduction to Electro-Optic Devices, Academic Press, NY, pp. 54 (1974).

4. Hinds International, Portland, OR.

5. Xtron Inc., Hayward, CA.

6. Fraatz, R. J., Rahe, W. H. and Allen, F. S. Computers and Chemistry, 5, 101 (1981).

7. Rahe, W. H., Fraatz, R. J., Sun, L. K. and Allen, F. S. Rev. Sci. Instr., xxx.

MORE INFORMATION FROM YOUR CIRCULAR DICHROISM

W. Curtis Johnson, Jr.

Department of Biochemistry and Biophysics
Oregon State University
Corvallis, Oregon 97331

INTRODUCTION

The purpose of this meeting is to discuss how circularly polarized synchrotron radiation could be used to learn more about molecules. Synchrotron radiation opens the possibility for high fluxes of circularly polarized light at very high energy. Thus the exciting new techniques such as circular intensity differential scattering, and circular differential microscopy would benefit greatly from a synchrotron ring generating circular polarized light. I believe such radiation will prove equally useful for that standard technique, circular dichroism of electronic absorption bands. Time constants of 10 to 60 seconds are the norm on conventional instrumentation so that it may take a number of hours to scan a CD spectrum. The large increase in light intensity from a synchrotron source allows shorter time constants and more rapid scanning, and smaller spectral slit widths for increased resolution. More rapid scanning offers the potential of doing kinetic studies on changes in secondary structure for biopolymers.

Synchrotron sources can provide high intensity radiation between 200 and 100 nm where the bulk of the electronic absorption bands occur. While this is a very modest energy for a synchrotron source, it is an extremely important spectral region for which no intense conventional sources are available. Patricia Snyder and John Sutherland are pioneering the use of synchrotron radiation in this wavelength region and have already discussed how it shortens scanning time and can be used to increase spectral resolution. A number of workers have built CD instrumentation for the vacuum UV region around commercial sources. Gene Stevens has already discussed his work on peptides and saccharides. I would like to use spectra that we have measured on our vacuum UV CD instrument to show why this spectral region is important. The take-home message is simple: Measuring more CD bands means more information about the molecule and a greater sensitivity to secondary structure.

Fig. 1 shows CD spectra that we measured for L-prolyl-L-proline diketopiperazine in three different solvents. With the very transparent solvent 1,1,1,3,3,3-hexafluoroisopropanol (HFIP), the CD spectrum could be measured to 135 nm and displays six CD bands as well as a shoulder at about 190 nm.[1] Earlier spectra measured on commercial instrumentation cut off at 190 nm.[2] Thus we have doubled the amount of CD information available for this molecule by making measurements into the vacuum UV. The intense bands measured at about 175, 160, and 140 nm have been assigned to $\sigma\sigma*$ transitions of the backbone. The biological solvent, water, is not as transparent as HFIP, but measurements can still be made to 165 nm if extremely short pathlengths are used, as Fig. 1 also shows.

POLYSACCHARIDES

Unsubstituted carbohydrates do not have any absorption bands above 200 nm and thus cannot readily be studied on commercial CD instrumentation. We have studied the CD of a number of monosaccharides in the

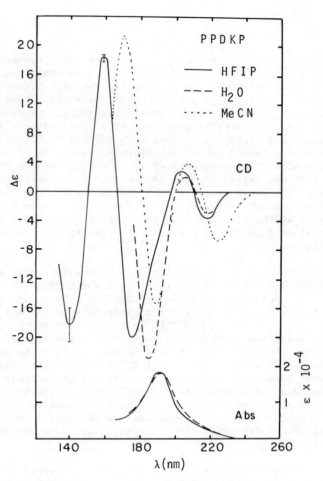

Fig. 1. The CD and absorption spectra of L-prolyl-L-proline diketopiperazine in HFIP (———), H_2O (------), and acetonitrile (••••••) from ref. 1.

vacuum UV and related the observed bands to the structure of the
molecules.[3-5] Fig. 2 shows the spectra measured for the anomers of D-
galactopyranose.[4]

    We have also done an extensive study of the polysaccharide amylose
to answer the simple question: Is amylose helical in solution? Nearly
everyone agrees that molecules such as iodine and n-butanol that bind to
amylose complex into an interior channel of a helical structure. However, it was not clear whether or not amylose is helical in the absence of
such a complexing agent. We measured the CD of amylose in aqueous solu-

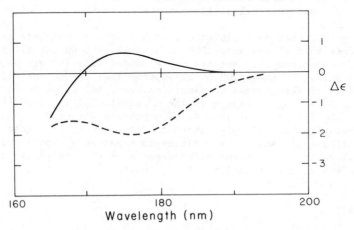

Fig. 2.   The CD spectra of D-galactopyranose as the α-anomer
          (————) and the β-anomer (------), redrawn from ref. 4.

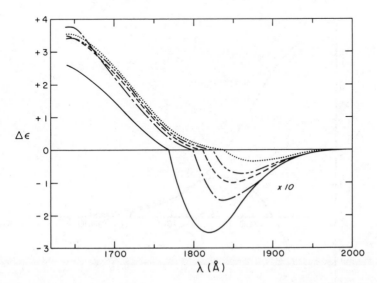

Fig. 3.   The CD spectra of maltose (••••••), maltotriose (- • • -),
          maltotetraose (------), maltohexose (- • -), and amylose
          (————) in aqueous solution at 10°C, from ref. 6.

tion (Fig. 3) and found it to be identical to the CD of amylose in the presence of n-butanol.[6] This is strong evidence that amylose is helical in aqueous solution. Fig. 4 compares the CD spectra of amylose and cyclo-hexylamylose, which are chromophorically identical.[6] Cyclohexylamylose and helical amylose are believed to have very similar secondary structures except that cyclohexylamylose has zero pitch and thus no helical chira-lity. We believe that the difference in CD between cyclohexylamylose and amylose results from the helicity of amylose in aqueous solution. Fig. 3 also shows that there is no sudden change with increasing chainlength which would indicate the sudden formation of a helical structure. This indicates a helicity for all of the oligomers regardless of chainlength which is dependent upon steric considerations. These results are con-sistent with conformal calculations based on steric constraints.[7-9]

The polysaccharide hyaluronic acid has both carboxylate and amide chromophores that absorb above 200 nm. Still, the amount of information is greatly increased when spectra are extended into the vacuum UV region, as Fig. 5 shows.[10] Above 200 nm it appears that the CD decreases as alco-hol is added to the aqueous solvent. However, below 200 nm it is clear that there is a large increase in the CD as alcohol is added,[11] which indicates the formation of an ordered secondary structure. Paul Staskus is now investigating this phenomenon to determine what is going on. His work on oligomers[10] shows the chainlength dependence expected for a heli-cal structure and the concentration dependence of this change indicates the secondary structure involves two polymers.

NUCLEIC ACIDS

Because of its sensitivity to secondary structure, CD has been used extensively to study DNA. However, above 200 nm CD is nearly insensitive to heat denaturation of DNA as Fig. 6 demonstrates.[12] Here we have a technique that is particularly sensitive to changes in secondary structure

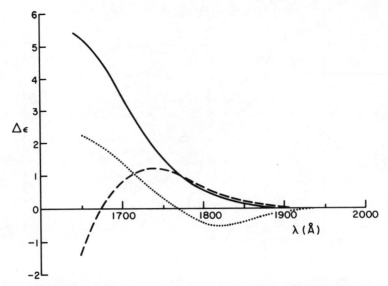

Fig. 4. The CD spectra of amylose (••••••), cyclohexoamylose
        (————), and methyl-α-D-glucoside (------) from ref. 6.

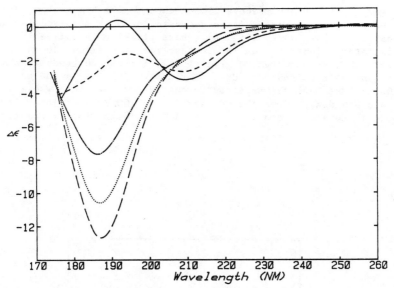

Fig. 5. The CD of hyaluronic acid in aqueous solution at 20°C, pH 2.8, with 0% (————), 12.4% (------), 13.8% (- • -), 18.3% (••••••), and 46.4% (— — —) 2,2,2-trifluoroethanol on a volume-volume basis (Staskus and Johnson, unpublished research).

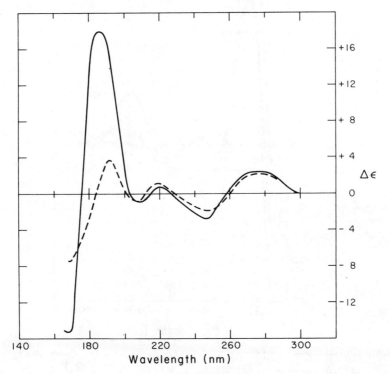

Fig. 6. The CD of E. coli DNA in D_2O with 1 mM NaF in its native form at 20°C (————) and heat denatured at 60°C (------), redrawn from ref. 12.

and yet very little happens above 200 nm when native DNA is heat denatured. However, the expected large decrease in CD is observed in the
bands measured below 200 nm (Fig. 6). Base stacking in native DNA results
in strong interactions between the intense ππ* transitions in the plane of
the bases. This is expected to give rise to intense CD bands as observed
for many synthetic polynucleotides. For native DNA, it is the CD bands
below 200 nm that have the expected intensity and the bands above 200 nm
are anomalously weak. Thus the CD bands below 200 nm lose intensity as
native DNA is denatured and there is no longer the strong interaction that

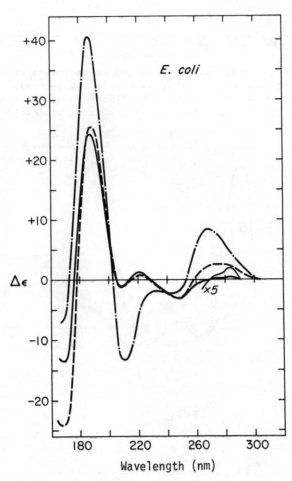

Fig. 7.   The CD of E. coli DNA as the 10.4 B-form in 10 mM $Na_{3/2}H_{3/2}PO_4$,
pH 7.0 (------); as the 10.2 B-form in 6 M NH_4F, 10 mM
$Na_{3/2}H_{3/2}PO_4$, pH 7.0 (———), and as the A-form in 80% tri-
fluororoethanol, 0.667 mM $Na_{3/2}H_{3/2}PO_4$ (- • -) from ref. 13.

occurs when the bases are stacked. It is not known why the CD bands above
200 nm are so weak for native DNA, but it could well be that the seven $\pi\pi^*$
transitions produce intense CD bands[12] that nearly cancel, giving a shape
and intensity that is accidentally similar to the CD bands for denatured
DNA.

Fig. 7 shows the CD spectra for E. coli DNA in the A-form, the B-form
with 10.4 base pairs per turn, and the B-form with 10.2 base pairs per
turn.[13] The CD of A-form DNA is clearly different from the CD spectra for
the two B-forms. We would not expect much difference in base-base interaction between 10.4 and 10.2 base pairs per turn, and this expectation is
borne out by the bands below 200 nm which are very similar in shape and
intensity. However, this small change in conformation can be monitored by
the 275 nm band, presumably because cancellations between the CD bands in
this region are slightly altered.

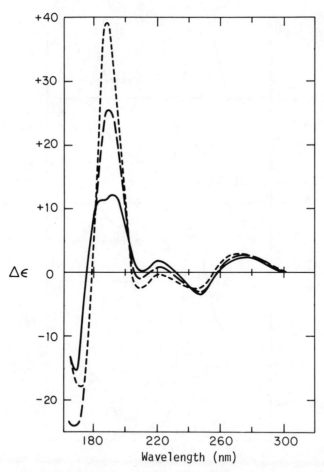

Fig. 8.   The CD spectrum of M. luteus DNA (------), E. coli DNA (— —)
and Cl. perfringens DNA (————), as the 10.4 B-form in 10 mM
$Na_{3/2}H_{3/2}PO_4$, pH 7.0, redrawn from ref. 13.

Since the CD bands below 200 nm are sensitive to base-base interactions, we would expect these bands to be sensitive to the GC content of various DNAs. The expected results are found in Fig. 8 where M. luteus with a high GC content has particularly intense CD bands in the vacuum UV while E. coli DNA with a 50% GC content has a lower intensity at 185 nm and Cl. perfringens with a low GC content has even a less intense 185 nm band and shows a double maximum as well.[13]

PROTEINS

The CD spectra of model polypeptides can be extended far into the

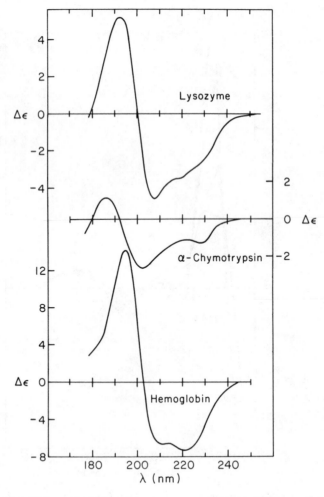

Fig. 9. The CD spectra for lysozyme, α-chymotrypsin, and hemoglobin, redrawn from ref. 14.

vacuum UV when they are studied in films or particularly transparent
solvents. We have measured the CD of a number of natural proteins in
aqueous solution,[14] and here the cutoff is 178 nm since we used a commer-
cially available cell with a 50 μ pathlength. Fig. 9 demonstrates that CD
is indeed sensitive to differences in secondary structure for various proteins and that the amount of information is roughly doubled by measuring
CD spectra to 178 nm rather than stopping at 200 nm.

Singular value decomposition (SVD)[15] can be used to show that CD
spectra of proteins measured to 178 nm contain five pieces of information
while CD spectra measured to 200 nm contain only two pieces of information.[14] Even if one lumps all the different types of β-turns together,
there are at least five parameters that contribute to the CD spectrum of a
protein: α-helix (H), antiparallel β-sheet (A), parallel β-sheet (P), β-
turns (T), and other structures not included in the first four categories
(O). Thus it is quite clear that the two equations provided by CD spectra
of proteins measured to 200 nm and the third equation which constrains the
sum of secondary structures to be 100%, do not provide enough information
to solve for five unknowns. We can, however, solve for these five
unknowns by using the five equations provided by the CD spectra of proteins
measured to 178 nm. We have used the five significant basis CD spectra
generated by applying SVD to the CD spectra of 16 proteins with known
secondary structure to investigate proteins with unknown structure.[14]
Fig. 10 shows the five significant basis spectra which correspond to a
mixture of secondary structures (Table 1), just as the CD spectrum of a
protein corresponds to a mixture to secondary structures. We do not
constrain the sum of secondary structures for the unknown protein to equal
100%, but use this fact to judge the reliability of our analysis.

Fig. 11 shows the CD spectrum of acetylcholinesterase[16] which has
been analyzed to contain 34% α-helix, 13% antiparallel β-sheet, 8%
parallel β-sheet, and 16% β-turns, and 26% other, for a total of 97%. The
CD spectrum of interferon,[17] shown in Fig. 12, analyzes as 59% α-helix,
16% antiparallel β-sheet, no parallel β-sheet, 18% β-turns, and 13% other
structure, for a total of 106%.

Table 1. Secondary Structure Corresponding to the Five Significant
          Basis CD Spectra

Basis	H	A	P	T	O
I	0.65	0.04	0.04	0.09	0.18
II	0.24	-0.09	0.21	0.14	0.31
III	0.13	0.25	0.03	0.18	0.04
IV	-0.28	0.51	0.11	-0.01	-0.09
V	0.07	-0.27	0.12	0.09	0.44

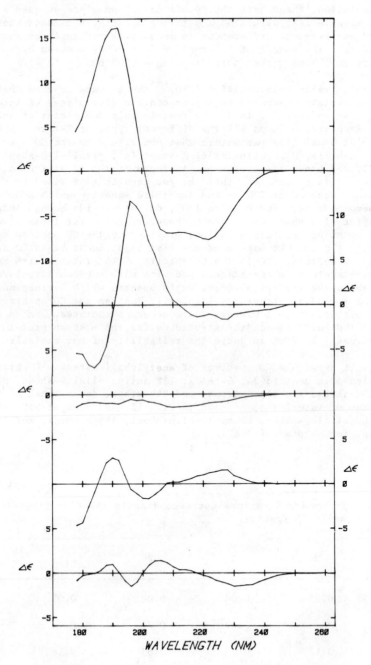

Fig. 10. The five significant CD basis spectra for analyzing the CD
spectrum of a protein for secondary structure, redrawn from
ref. 14.

CONCLUSION

The extra information provided by extending CD spectra into the vacuum UV region provides extra sensitivity for investigating the secondary structure of all types of biological molecules. This extra information is mandatory if CD spectra of proteins are to be used to determine their component secondary structures. The noise on a CD spectrum depends on the amount of light collected during the measurement and conventional sources for the vacuum UV are weak enough that noise levels are high and the instruments always working at their limits of reliability. Synchrotron radiation provides the high light intensity for this wavelength region that is necessary for the rapid scanning of accurate noise-free CD spectra.

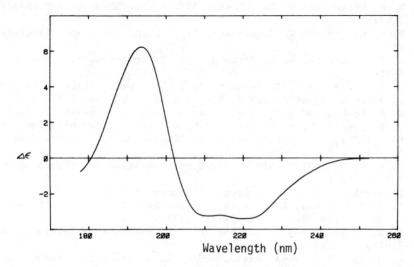

Fig. 11. The CD of acetylcholinesterase (Manavalan, Johnson, and Taylor, unpublished research).

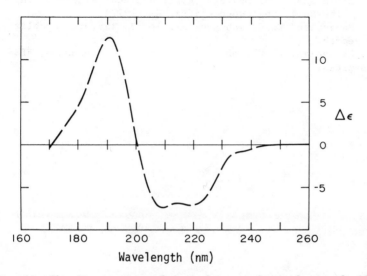

Fig. 12. The CD spectrum of interferon redrawn from ref. 17.

ACKNOWLEDGMENT

    This work was supported by National Science Foundation grant
PCM80-21210 from the Biophysics Section and Public Health Service
grant GM21479 from the Institute of General Medical Sciences.

REFERENCES

1. R. L. Bowman, M. Kellerman, and W. C. Johnson, Jr., Biopolymers
    22:1045-2070 (1983).
2. T. M. Hooker, Jr., P. M. Bayley, W. Radding, and J. A. Schellman,
    Biopolymers 13:549-566 (1974).
3. R. G. Nelson and W. C. Johnson, Jr., J. Am. Chem. Soc. 94:3343-3345
    (1972).
4. R. G. Nelson and W. C. Johnson, Jr., J. Am. Chem. Soc. 98:4290-4295
    (1976).
5. R. G. Nelson and W. C. Johnson, Jr., J. Am. Chem. Soc. 98:4296-4301
    (1976).
6. D. G. Lewis and W. C. Johnson, Jr., Biopolymers 17:1439-1449 (1978).
7. P. Zugenmaier and A. Sarko, Biopolymers 12:435-444 (1973).
8. K. G. Goebel and D. A. Brant, Macromolecules 3:634-643 (1970);
    C. V. Goebel, W. L. Dimpfl, and D. A. Brant, Macromolecules 3:634-
    643 (1970); D. A. Brant and W. L. Dimpfl, Macromolecules 3:644-654
    (1970).
9. V. S. R. Rao, N. Yathindra and P. R. Sundararajan, Biopolymers 8:
    325-333 (1969).
10. P. Staskus and W. C. Johnson, Jr., unpublished research.
11. L. A. Buffington, E. S. Pysh, B. Chakrabarti, and E. A. Balazs,
    J. Amer. Chem. Soc. 99:1730-1734 (1977).
12. C. A. Sprecher and W. C. Johnson, Jr., Biopolymers 16:2243-2264
    (1977).
13. C. A. Sprecher, W. A. Baase, and W. C. Johnson, Jr., Biopolymers
    18:1009-1019 (1979).
14. J. P. Hennessey, Jr. and W. C. Johnson, Jr., Biochemistry
    20:1085-1094 (1981).
15. B. Noble and J. W. Daniel, "Applied Linear Algebra," 2nd Edition,
    Prentice-Hall, Englewood Cliffs, NJ (1977).
16. P. Manavalan, W. C. Johnson, Jr. and P. Taylor, unpublished
    research.
17. P. Manavalan, W. C. Johnson, Jr. and P. D. Johnston, FEBS Letters,
    in press.

CIRCULAR DIFFERENTIAL MICROSCOPY

Marcos F. Maestre*, David Keller**, Carlos Bustamante**,
and Ignacio Tinoco, Jr.***

*Biology and Medicine Division
Lawrence Berkeley Laboratory
        and
**Chemistry Department
The University of New Mexico
Albuquerque, New Mexico
        and
***Chemistry Department
University of California
Berkeley, California

1. INTRODUCTION

In this chapter we will describe the historical development
of the theory of differential imaging, and the invention of the
circular differential imaging microscope. The technique is shown
to be a logical extension of the research on the interaction of
circularly polarized light with structures whose dimensions are
arbitrary with respect to the wavelength of light, (Tinoco et
al., 1980).

The polarized light technique used initially to study these
biological structures was circular dichroism. Circular dichroism
is the difference in the extinction coefficient of matter when
illuminated by left and right circularly polarized light, and is
defined as:

$$\Delta\varepsilon = \varepsilon_L - \varepsilon_R,$$

where $\varepsilon_{L,R}$ is the specific extinction due to left and
circularly polarized light, respectively.

When circular dichroism measurements were done on large
biological structures, (i.e. bacteriophages), signals were

obtained in regions of the spectrum which did not correspond to the intrinsic absorbance bands of the material, (Maestre et al., 1971). These anomalous CD signals, outside the absorption region of the spectrum, are now interpreted to be a manifestation of a differential scattering cross section term (Bustamante, et al., 1983), which appears as a component of the extinction coefficient as described by the relation:

$$\epsilon_{L,R} = a_{L,R} + s_{L,R}$$

where $a_{L,R}$ are the intrinsic absorptions of the chromophores in the wavelength region specific to the material being measured, and $s_{L,R}$ are the scattering cross sections for incident left (L) and right (R) circularly polarized light. Outside the absorbance region $a_{L,R} = 0$, and the only term left in the above relation is that due to the differential scattering cross section $s_{L,R}$, (Bustamante, et al., 1983).

The spatial distribution of the intensities as a function of the scattering angle is described by a related quantity, the circular intensity differential scattering (CIDS), which characterizes the preferential scattering of right and left circularly polarized light by the sample at a given direction in space and is defined by;

$$CIDS = \frac{I_L - I_R}{I_L + I_R}$$

where $I_{L,R}$ are the intensities of the incident left and right circularly polarized light, respectively, (Tinoco et al., 1980).

A clear example of a spectrum of a biological structure that shows the influence of circular intensity differential scattering, (CIDS), on the CD, is presented in Fig. 1. The graph shows the CD spectrum measured in a standard circular dichrograph, a Cary 60, of a suspension of the helical sperm of the mediterranean octopus Eledone cirrhosa. The significant part of this CD spectrum is that there are strong signals in the wavelength regions above 300 nm, outside of the absorbance bands of the chromophores of nucleic acids and protein components of the sperm. These signals are due to the differential scattering of light by the sperm, a structure of remarkable complexity, being a left handed helix approximately 43 μm long, with a pitch of 1 μm and a radius of 0.65 μm, shown in Fig. 2 (Maestre et al., 1982a).

In the CD spectra displayed in Fig. 1 there are four curves which were taken by methods in which different amounts of correction are applied to the scattering components. These

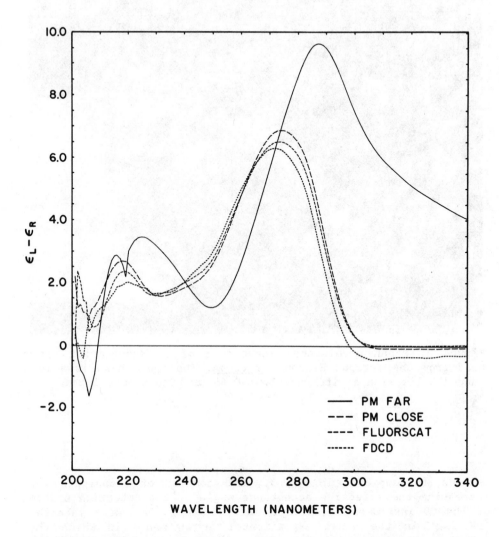

Figure 1: The circular dichroism spectra in arbitrary units of
E. cirrhosa sperm heads, measured by techniques with different
collection angles of scattered light. 'PM far' is the
conventional configuration, 'PM close' has the photomultiplier
detector close to the sample cuvette to collect the main beam
plus the forward scattered light. The Fluorscat and FDCD methods
are described in the text.

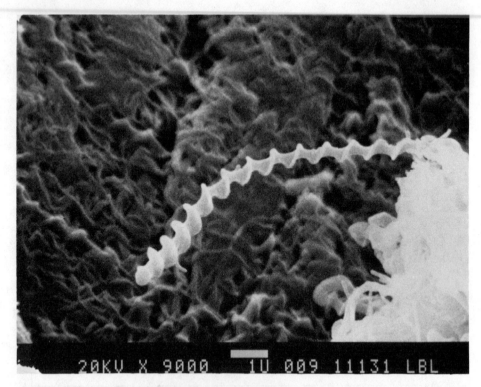

Figure 2: Scanning electron micrograph of a freeze-dried sperm head from the octopus <u>Eledone cirrhosa</u>. The sperm head is a left handed helix with a pitch of about 0.65 μm. Scale bar, 1 μm.

methods are depicted in Fig 3., and consist of techniques for increasing the effective acceptance angle of the detection system of the CD instrument, (Reich et al., 1980). The curve labelled 'PM far' is the usual experimental arrangement, in which the detector measures mainly the transmitted beam, and some of the light scattered close to the this beam is also measured. In this arrangement the CD signal is strong at long wavelength, where the components of the nuclei (DNA and protein) do not absorb light indicating a large contribution from differential scattering. This differential scattering is eliminated for the most part by bringing the photomultiplier close to the sample cuvette thereby increasing the angle of acceptance of the scattered light, (Fig. 1, curve labelled 'PM close'). Further small corrections were obtained by use of the Fluorscat cuvette, and the use of fluorescence detected circular dichroism, (FDCD), techniques as described in Fig. 3. The Fluorscat cuvette is a double chambered cell, in which the outer chamber is filled with a fluorescent solution whose absorbance bands are broad and lie within the region of the wavelength spectrum of interest, (Dorman and Maestre, 1973). This solution captures all the transmitted beam plus much of the scattered light, (i.e. that which is not scattered back towards the incident beam). The detector measures the fluorescent light since there is total absorption of the incident beam by the fluorescent solution.

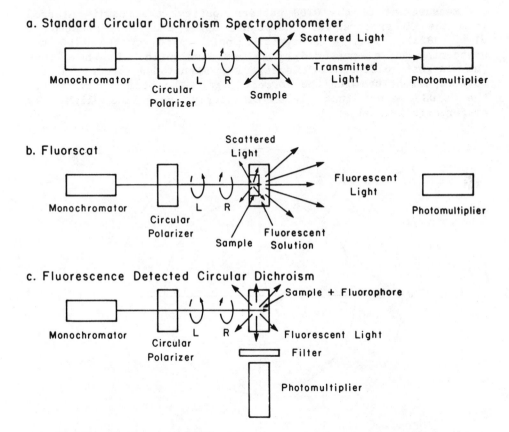

Figure 3: Various correction techniques for scattering contributions to CD. (a) Ordinary CD spectrophotometers have a narrow acceptance angle so cannot correct for differential scattering. (b) Fluorscat cuvettes will recover a large fraction of the scattered light by surrounding the inner cuvette with an outer cuvette containing a fluorescer whose spectral absorbance correspond to the wavelength region of interest. (c) FDCD is used to correct for scattering by surrounding the particle in suspension by a non-optically active solvent that will convert transmitted and scattered light into fluorescent light for detection.

The FDCD technique involves the use of a fluorescer molecule that is not optically active, and does not interact chemically or bind to the particle being measured. Each scattering particle is now surrounded by a cloud of fluorescent molecules which measure all the light that is not absorbed, including light scattered backwards along the incoming light beam. Considering the minor corrections to the measured CD by fluorscat and FDCD techniques (Fig. 1), we conclude that the sperm heads scatter light differentially mainly in the forward direction.

Measurement of the CIDS pattern confirms the result inferred from the CD spectra vs. acceptance angle studies, (Maestre et al., 1982a). Fig. 4 presents a polar plot of the CIDS vs. scattering angle, measured at 442 nm wavelength. It shows that the largest differential scattering magnitudes occur in the forward direction with the correct sign. The positive value for the CIDS means that left circularly polarized light is preferentially scattered.

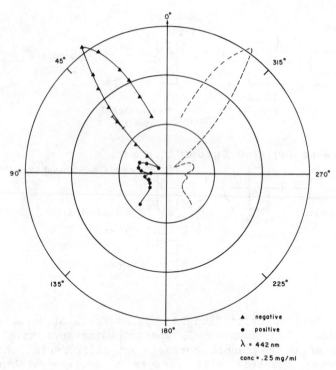

Figure 4: Plots of CIDS versus scattering angle for _Eledone cirrhosa_ sperm heads. The direction of the incident beam is given by the arrow, (wavelength of 442 nm). The triangles correspond to negative values and the circles to positive values of CIDS. Positive values means left circularly polarized light is preferentially scattered.

Theoretical analysis of helical models (Bustamante et al.,
1980, 1981, 1982, 1983, 1984a) have shown that the CIDS patterns
are most sensitive to the helical parameters. By accurately
measuring the CIDS patterns as a function of wavelength,
information about the chiral organization of the scattering
structures can be obtained, (Tinoco, et al., 1982, Bustamante et
al., 1984).

A conclusion from extensive measurements by use of circularly
polarized light of large intact biological materials, is that
structural information can be extracted both from the direct
measurement of the true differential absorbance $a_L - a_R$, and
the differential scattering cross sections, $s_L - s_R$ or from
the CIDS patterns, (Tinoco et al., 1983). The differential
absorbance is a function of the secondary chiral organization,
and the differential scattering cross section is a function of
tertiary and higher, chiral order. In the next section we will
discuss how these concepts are applied to the measurement of
single isolated intact biological structures by microscopic
polarization techniques.

2. THE CIRCULAR DICHROISM MICROSCOPE

CD spectra are usually measured on solutions or suspensions
of the materials to be studied, and the signals are the average
of the population. In most cases, the population average well
represents the properties of each of the members of the group
because of population homogeneity. This is not true for a
population such as living mammalian cells, since there is a whole
distribution of cells in different stages of their life cycle.
The study of the optical properties of such systems requires the
measurement of each cell individually. The development of a
microscope capable of measuring the CD of each microscopic
particle, would permit the study of the differential absorbance,
$a_L - a_R$, and the differential scattering cross section,
$s_L - s_R$, along the cell life cycle.

3. INSTRUMENT DESIGN

The CD microspectrophotometer is a combination of two
existing instruments, a Cary 60 dichrograph, and a Zeiss
photometer microscope (PM-3) with UV transparent objectives,
(Maestre and Katz, 1982b). The Zeiss objectives are strain free
and permit measurements down to 200 nm wavelength. The use of
objectives with differing numerical aperture such as 32 X
power,(N.A.=0.4, glycerol imm.), and 100 X power, (N.A.=1.25,
glycerol imm.), permits a measurement of the differential
scattering cross-sections of the microscopic object as a function
of the acceptance angle of the lenses. This distinction is
important because the two different objectives capture different

amounts of the differentially scattered light. The acceptance angle of the 32 X power objective is 31° degrees, and that of the 100 X power is 116°. The higher power objective will capture more scattered light than the lower power objective. The variation of the acceptance angle permits some degree of determination of the differential scattering contribution in the forward directions of scattering regions. This forward differential scattering component can be used to extract information about the long range chiral organization of biological structures under the microscope.

The CD microscope has been used to study the CD/CIDS spectra of Chinese Hamster tissue culture cells, (CHO), as a function of the life cycle of the cells, (Salzman and Maestre, 1981) and (Maestre, Salzman, Bustamante and Tobey, ms. in preparation). It can be shown that there are distinct spectra associated with the different states of the cell in their life cycle, and that correlations exist with the amount of DNA in the cell as monitored by absorbance at 260 nm (measured by the CD microscope).

This is shown in Fig 5. which presents averages computed for the CD spectra taken with the narrow acceptance angle objective of single cells in the M, G1, S, and G2 phases. At the narrow acceptance angle objective the contribution due to differential scattering is maximum for the instrument. Changes in the differential scattering components, if any, are easily detected in this instrument mode as the cell moves through the life cycle. Fig. 5 shows scattering contributions in the regions above 300 nm, where the cells have no optical absorbance. The curve that displays the maximum scattering contributions is the spectrum labeled M, in which the chromatin is presumably condensed into the chromosomal organization.

Data measured with the 100x objective shows the scattering components to be minimized, as seen by the reduced CD "tails" above 300 nm, (Maestre, Salzman, Bustamante and Tobey, ms. in preparation) There are alterations in all sections of the CD spectrum as low as 220 nm. The data indicate that the scattering components affect particularly the lower regions of the spectrum towards the protein bands. A crude measure of the spatial distribution of the scattering can be computed, by subtracting the data for each cell phase, obtained with the 100X objective, from data obtained by 32X objective measurements. These difference CD spectra carry information on the large chiral structures in the cells. The wavelength behavior of these curves also point out which region of the complete CD spectrum provides the maximum contributions to the scattering pattern.

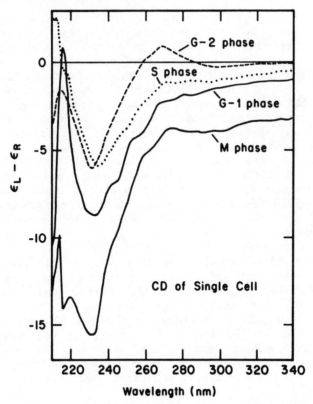

Figure 5: CD/CIDS microscope spectra for single Chinese Hamster
Ovary cells as a function of life cycle. These data were taken
with a 32x power objective which correct little for the
differential scattering contributions. The CIDS scattering
components are manifested in the spectra as apparent CD in
wavelength regions higher than 300 nm.

4. THE DIFFERENTIAL IMAGING MICROSCOPE

In the CD/CIDS microscope, the objective lens in the
instrument only serves to collect the transmitted light through
the microscopic object plus some fraction of the scattered light,
depending on the aperture of the lens. The collected light is
not used to form an image, and instead is measured by the
standard photomultiplier detector of the CD instrument to give
spectral curves.

If an imaging system (a lens or a microscope) is then placed
between the chiral sample and the detector, an image can be

formed.   The image produced by left circularly polarized
illumination will be different from the image produced when the
illumination is right circularly polarized if the object being
imaged is chiral and interacts preferentially with one of the
circular polarizations of light. The difference between these
two images can be called a circular differential image. With the
proper detection system and associated electronics, we can use
the imaging properties of the microscope to form a differential
image, with a spatial distribution of information along the
image.   The information in the image is now a point by point
measurement of the CD, in the case of transmitted light
measurement, or differential scattering contribution to the image
cross-section, in the case of dark-field illumination, (Keller
et al., 1985). The spatial resolution of the measurement is
limited by the lens resolving power, pixel geometry in the
detector, and available light intensity.

The information contained in the circular differential image
will not be the same information as that of the image produced
using unpolarized light.   In the latter case the optical
contrast   that distinguishes one feature of the sample from
another is provided by differences in the absorption and the
index of refraction of the various parts of the sample. In a
circular differential image the contrast is provided by
differences on the interaction of different parts of the sample
with left and right circularly polarized light.

It has been shown that circular differential scattering is
specially sensitive to the dimensions of the structure close to
the wavelength of the incident light, and application of CIDS
theory to images extend these results. Particularly instructive
is the behavior of the images as a function of wavelength of the
illuminating circularly polarized light.   Figure 6 shows two
helices, the smaller being ten times smaller in diameter and
pitch than the larger one. The helices will be illuminated by
circularly polarized light of varying wavelength and a
differential image produced by the scattered light (darkfield
illumination) will be computed by subtracting from the image
produced by right circularly polarized light that produced using
left circularly polarized light.   Fig. 7 shows the computed
image generated by light of wavelength comparable to the
dimensions of the large helix. Fig. 8 shows the image produced
by light of wavelength of dimension of the order of the small
helix. The difference images point to a characteristic property
of the technique that gives a zero differential image for those
objects that are not chiral in structure. These non-chiral
structures will no be seen by the instrument. Furthermore, by
varying the wavelength different sizes of chiral structures would
be emphasized and the sense of chirality of the object can be
determined by the signs associated with the image.

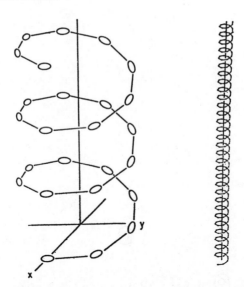

Figure 6: Scattering particle arrays in the shape of two helices, the smaller one having pitch and diameter in tenth in magnitude of the large helix. These helices will produce the differential images displayed in Figs. 7 and 8 for darkfield illumination of varying wavelength.

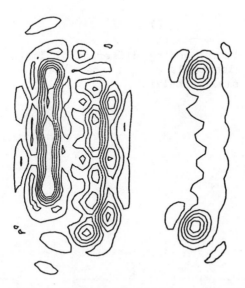

Figure 7: The computed differential image, (right – left), for the helices in Fig. 7 illuminated by light whose wavelength is of the order of the pitch and diameter of the large helix.

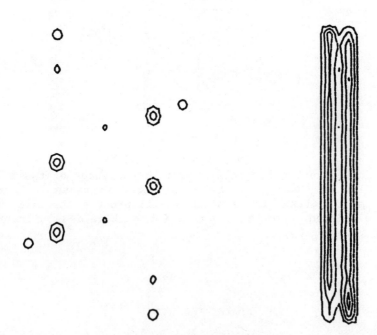

Figure 8:   The corresponding differential image for illuminating light whose wavelength is now of the dimension of the pitch and diameter of the smaller helix.

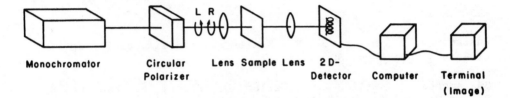

Circular Differential Microscopy

Figure 9: A schematic diagram of a circular differential imaging microscope. The intensity at each point in the image plane is measured when left and right circularly polarized light is incident on the sample. Only chiral objects will be evident in the image.

A differential microscope has been constructed following the design shown in Fig. 9, (Mickols, Embury, Maestre and Tinoco, ms. in preparation). Preliminary studies using this instrument on the polymer formation of hemoglobin S in the sickling of intact red blood cells have shown that the technique can give valuable information on the spatial distribution and concentration of the polymer on both sickled and irreversibly sickled cells in situ. The technique shows promise as a new way for measuring the spatial distribution of chirality inside microscopic biological materials. This information can then be used to construct possible models of the biological organization.

ACKNOWLEDGEMENTS

This work was supported in part by National Institutes of Health Grants AI 08247 (M.F.M), GM 32543 (C.B.), GM 10840 (I.T.), a Searle scholarship awarded to C.B. (1984), and by the U.S. Dept. of Energy, Office of Energy Research Contract DE–AT03–82–ER60090. Partial support was also provided by a grant from Research Corporation (C.B.).

REFERENCES

Bustamante, C., Maestre, M.F., Keller, D. and Tinoco, I., Jr. (1984a) J.Chem.Phys. 80, 4817.

Bustamante, C., Wills, K.S., Keller, D., Samori, B., Maestre, M. P. and Tinoco, I., Jr., (1984b) Mol. Cryst. Liq. Cryst. 111, 79-102.

Bustamante, C., Tinoco, I., Jr. and Maestre, M.F. (1983) Proc. Natl. Acad. Sci. U.S.A., 80, 3568-3572.

Bustamante C., Tinoco, I., Jr. and Maestre, M.F.(1982) J.Chem.Phys. 76, 3440.

Bustamante, C., Tinoco, I., Jr. and Maestre, M.F.(1981) J.Chem.Phys. 74: ,4839

Bustamante, C., Maestre, M.F. and Tinoco, I., Jr.(1980) J.Chem.Phys. 73:,6046

Dorman, B.P. and Maestre, M. F. (1973) Proc. Nat. Acad. Sci. USA 70 :255-59

Keller, D., Bustamante, C., Maestre, M.F. and Tinoco, I., Jr. (1985), Proc. Nat. Acad. U.S.A., in press.

Maestre, M.F. and Katz, J. (1982b) Biopolymers 21:1899.

Maestre, M.F., Bustamante, C., Hayes, T.L., Subirana, J. A., Tinoco, I., Jr. (1982a) Nature, 298: 773-774

Maestre, M.F., Gray, D. M. and Cook, R.B. (1971a) Biopolymers 10 : 2537

Reich, C., Maestre, M.F, Edmonson, S. and Gray, D.M.(1980) Biochemistry 19: 5208

Salzman, G., and Maestre, M.F. (1981) Cytometry Vol. 2: 125

Tinoco,I. Jr.,Bustamante, C. and Maestre,M.F.(1980) Ann. Rev. Biophys. Bioeng 9:107-41

Tinoco, I., Jr., Bustamante, C. and Maestre, M. F. (1982) in "Structural Molecular Biology" Eds: Davies, D.B., Saenger, W. and Danyluk, S.S., Plenum.

Tinoco, I. Jr., Maestre, M.F. and Bustamante, C. (1983) Trends in Biochem. Sci. 8:41-44.

CIRCULAR DICHROISM OF LARGE ORIENTED HELICES:

A FREE ELECTRON ON A HELIX

Dexter S. Moore

Department of Chemistry
Howard University
Washington, D.C. 20059

INTRODUCTION

An electron free to move on a helical path is a simple optically active system; the optical properties of which can be calculated exactly.
Such a helical system was discussed in the small molecule (or dipole or
Rosenfeld) approximation by Tinoco and Woody (1964). The wavefunctions
and energies necessary to calculate the optical properties of the helix
are identical to those for a particle in a one-dimensional box. Thus,
this simple optically active system offers the possibility of furnishing
insights into the origins of the optical properties of many chiral systems. For example, it is of interest to understand the origins of the
interesting circular dichroism (CD) bands often associated with molecular
aggregates such as cholesteric liquid crystals, chromosomes, nucleic acid
in various condensed states, viruses, etc. These chiral systems generally have spatial dimensions that are large compared to the wavelength of
the incident light giving rise to the CD bands. Thus, the usual Rosenfeld approximation to the optical properties of such systems will, in
general, be inadequate.

In the Rosenfeld approximation, the CD and optical rotation of light
by randomly oriented chiral systems have been shown (Rosenfeld, 1928) to
be obtainable from the following equations:

$$\theta = (8\pi N\nu^2/3hc) \; \Sigma \; R_{oa}/(\nu_a^2 - \nu^2),$$

$$R_{oa} = Im \; \mu_{oa} \cdot m_{ao},$$

$$\mu_{oa} = \int \Psi_o^* \; \mu \; \Psi_a \; d\tau; \quad m_{ao} = \int \Psi_a^* \; m \; \Psi_o \; d\tau.$$

The angle θ is the rotation (radians/cm) for N molecules/cm^3 at light
frequency ν. Roa is the rotational strength for a transition from ground
state, o, to excited state, a, with frequency ν_a. The rotational strength
is equal to the imaginary part of the dot product of the transition electric dipole moment, μ_{oa}, and transition magnetic dipole moment, m_{ao}. The
CD is directly proportional to the rotational strength. The expression
for Roa was derived for molecular dimensions small compared to the wavelength of light by expanding the exponentials of the transition probability integrals (that appear in the correct vector potential of the

radiation in the interaction Hamiltonian) and retaining the dipole terms, μ_{oa} and m_{ao}. Thus, as the dimensions of a chiral system become large relative to the wavelength of light leading to the optically active transition the Rosenfeld approximation should become inadequate.

For the CD and optical rotation of large chiral systems, it is often necessary to retain the exponentials in correct form rather than making the Rosenfeld approximation. For the electron on a helix system, Moore and Tinoco (1980) derived equations that are applicable to helices of any dimensions. The equations were found to give results in good agreement with measured rotations of microwaves by oriented wire helices. In the following, some results are presented for the CD of oriented helices calculated as a function of variations in the dimensions of the free electron helical system.

THE MODEL

To calculate the CD of the oriented helices, one must specify the direction of incidence and the state of polarization of the light as well as the helical dimensions. In the following, results of CD calculations are presented for light incident perpendicular to the helix axis and parallel to this axis, Fig. 1. Light incident along the X, Y, and Z directions of the helix (Fig. 1) is therefore considered. Left and right circularly polarized light with unit vectors (as defined below) were employed. In order to specify the helix, the critical parameters are the pitch ($2\pi b$) and the radius (a), Fig. 1. For the sake of simplicity, only right-handed, one-turn helices are considered in the calculations.

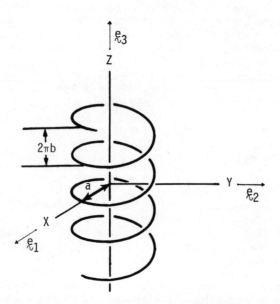

Fig. 1. A right-handed helix with pitch ($2\pi b$) and radius (a).

EQUATIONS

The CD for helices of arbitrary size can be calculated as shown previously (Moore and Tinoco, 1980). For example, for X-incident light the
CD is given by

$$E_L - E_R (\mathfrak{e}_1 \text{-incident}) = 3G(e\lambda_{nm}/2\pi\mu c)^2 \text{Im}(\mathfrak{e}_3 \cdot \mathfrak{T}_{nm} \mathfrak{T}_{nm}^* \cdot \mathfrak{e}_2 - \mathfrak{e}_2 \cdot \mathfrak{T}_{nm} \mathfrak{T}_{nm}^* \cdot \mathfrak{e}_3), \quad (1)$$

where,

E_L = absorptivity for left circularly polarized light with unit vector $\mathfrak{e}_L = (1/2)^{\frac{1}{2}}(\mathfrak{e}_2 - i\mathfrak{e}_3)$;

E_R = absorptivity for right circularly polarized light with unit vector $\mathfrak{e}_R = (1/2)^{\frac{1}{2}}(\mathfrak{e}_2 + i\mathfrak{e}_3)$;

$\mathfrak{e}_1$, $\mathfrak{e}_2$, $\mathfrak{e}_3$ = respective, x, y, z unit vectors, Fig. 1;

$G = 8\pi^3 N_o \lambda g(\lambda)/6909hc$; e, u = charge, electron mass; c = speed of
light; $g(\lambda)$ = band-shape function of unit area; N_o, h = Avoga-
dro's number, Plank's constant;

$\lambda_{nm} = 32\pi^2 \mu k^2(a^2 + b^2)c/h(m^2 - n^2) = hc/(E_m - E_n)$; n,m = ground, excited
state quantum numbers; E_n, E_m = energies of the states; k = num-
ber of helical turns;

$\mathfrak{T}_{nm} = \int \Psi_n^* \exp(i2\pi\mathfrak{e}_1 \cdot \mathfrak{r}/\lambda_{nm})\mathfrak{P}\Psi_m d\tau$, the transition probability integral; Ψ_n, Ψ_m = ground, excited state wavefunctions; r, P =
position, linear momentum operators.

To calculate the CD in the Rosenfeld approximation, the exponentials
appearing in the above transition probability integrals are expanded and
the dipole terms kept. The result is

$$E_L - E_R (\mathfrak{e}_1 \text{-incident}) = 6G(e\lambda_{nm}/2\pi\mu c)^2 (2\pi/\lambda_{nm}) (\mathfrak{e}_3 \cdot \mathfrak{r} \cdot \mathfrak{e}_1 \cdot (\mathfrak{r}\mathfrak{P}) \cdot \mathfrak{e}_2$$
$$- \mathfrak{e}_2 \cdot \mathfrak{r} \cdot \mathfrak{e}_1 \cdot (\mathfrak{r}\mathfrak{P}) \cdot \mathfrak{e}_3) . \quad (2)$$

Averaging Eq. (2) over all orientations results in the familiar Rosenfeld
approximation for randomly oriented free electron helices:

$$E_L - E_R = 4G \text{ Im}(\mu_{nm} \cdot \mathfrak{m}_{mn}); \quad \mathfrak{m}_{mn} = e/2\mu c \int \Psi_m^*(\mathfrak{r} \times \mathfrak{P})\Psi_n d\tau . \quad (3)$$

The corresponding equations for the average CD based on the transition
probability integrals, Eq. (1), are quite complicated. They have been
discussed by Balazs et al. (1976).

In performing the calculations it should be noted that the transition
wavelengths, λ_{nm}, and the helical parameters are not independent (see
above). In the following calculations, the transition wavelength is kept
approximately constant as the helical parameters (a and b) are varied, by
choosing (m-n) = constant and allowing m+n to vary. For x and y incident
light, the important parameter affecting the CD is the radius (a). For

Z incident light the important parameter is the pitch.

ENERGY LEVELS AND SELECTION RULES

     Shown in Fig. 2 is a schematic energy level diagram for some transi-
tions of the electron on a helix. N and M represent the highest occupied
and lowest unoccupied energy levels, respectively. There are two electrons per occupied level. For a typical calculation, the longest wavelength transition will be N→M, i.e., a quantum number difference (ΔQN)
of 1. Occurring at shorter wavelengths will be two transitions, N→M+1
and N-1→M, both with ΔQN = 2 and approximately the same transition wavelength. At still shorter wavelengths, there will be three transitions
N→M+2, N-1→M+1, and N-2→M all with ΔQN = 3 and approximately the same
transition wavelength. The same trends are found for even shorter wavelength transitions, e.g., there are 9 transitions with ΔQN = 9, all occurring at short wavelengths with essentially the same transition wavelength.
However, not all transitions will be allowed. For a helix of an integral
number of turns (k), the CD selection rules for allowed transitions are
shown in Table I. The correct selection rules obtained from the tran-

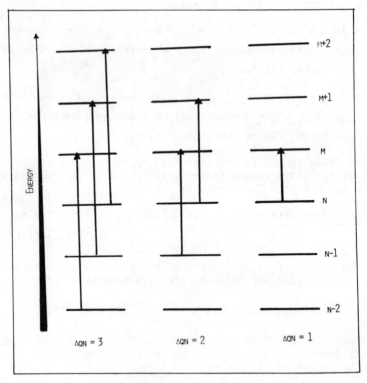

Fig. 2. A Schematic energy level diagram for the longest
          wavelength transitions of the electron on a helix.
          N = quantum number of highest occupied level; M =
          quantum number of lowest unoccupied level; ΔQN =
          excited level quantum number minus ground level
          quantum number.

Table 1. CD Selection Rules for an Electron on a Helix of
          k Turns

Light Incidence	Dipole Approximation	Transition Integral
X	m−n odd	m−n odd
Y	m−n odd; m+n = 2k	m−n odd; m+n = 2k, 6k, 10k...
Z	m−n odd; m+n = 2k	all transitions

sition integrals, Eq. (1), are observed (Table 1.) to differ somewhat from
those of the dipole (Rosenfeld) approximation.

RESULTS AND DISCUSSION

     Shown in Fig. 3 are CD spectra calculated for randomly oriented and
oriented helices that were right-handed with k = 1. The first three longest wavelength bands shown are composed of transitions with ΔQN of 1, 2,
and 3. The band shapes are arbitrary but their heights are proportional
to the calculated CD. The dimensions of the helices are such that the
transition integrals, Eq. (1), and the Rosenfeld approximation, Eqs. (2)
and (3), give identical results. The helical dimensions are set so that
$2\pi a = 2\pi b$. The lowest energy transition wavelength was chosen to be 120
nm. Thus, for the various transitions $2\pi a/\lambda_{nm} = 2\pi b/\lambda_{nm}$ have the values
indicated in parenthesis (Fig. 3). In the case of the CD for the oriented
helices (bottom, Fig. 3), all transitions are observed to be allowed for
the one-turn helices, except for X incident light (as indicated by the
selection rules, Table I) when m−n = even (i.e., ΔQN = 2).

     In Fig. 4, CD results are compared from the Rosenfeld approximation
and transition integrals (for ΔQN = 1) as a function of increasing helical dimensions. For convenience here, and in all following results,
$2\pi a = 2\pi b$. This is of no consequence, since the important parameter for
X and Y incidence is $\underline{a}$ and that for Z incidence is $\underline{b}$. The results shown
in Fig. 4 indicate the dimensions to wavelengths where the Rosenfeld
approximation becomes inadequate. For X and Y incidence (top, Fig. 4),
the transition integrals give CD that alternates in sign as the helical
dimensions become large relative to the transition wavelength but the
Rosenfeld approximation gives the same CD sign over all dimensions of the
helix. For Z incident light (bottom, Fig. 4), both the transition integrals and Rosenfeld approximation give negative CD for all helical dimensions. However, the CD from the Rosenfeld approximation increases in
negative magnitude with increasing helical dimensions while the CD from
the transition integrals show an overall magnitude decrease. In addition,
the transition integral CD show (bottom, Fig. 4) that there are ratios of
dimensions to wavelength where the CD goes essentially to zero but with
no change of sign as observed for X and Y incident light (top, Fig. 4).

     From the results shown for Z incident light (bottom, Fig. 4), it can
be concluded that for the longest wavelength transition (ΔQN = 1) the Rosenfeld approximation will always give the correct CD sign. Thus, for a
right-handed helix with light propagating parallel to the helix axis the
CD will always be negative for ΔQN = 1 and will always be positive for a

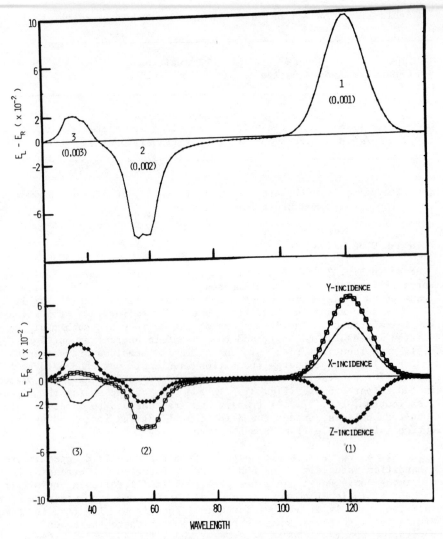

Fig. 3. Calculated CD spectra for randomly oriented (top) and oriented (bottom) helices of dimensions $2\pi a = 2\pi b = 0.120$ nm. The integers refer to ΔQN. The decimals in parentheses indicate the values of the ration of helix dimensions to transition wavelength. The band-shapes are arbitrary but their heights are proportional to the calculated CD.

left-handed helix. Thus, the handedness of optically active systems that correspond to the electron on a helix model could be determined by studying the CD of the longest wavelength transition for Z incident light; no knowledge of the helical dimensions would be needed.

For a one-turn helix, X-incident light and $\Delta QN = 2$, the electronic transitions are forbidden (Table 1) for all helical dimensions. For Y incident light and $\Delta QN = 2$, Fig. 5, the CD is found to be negative when the Rosenfeld approximation is valid (e.g., $2\pi a/\lambda_{nm} < 1.0$). As the he-

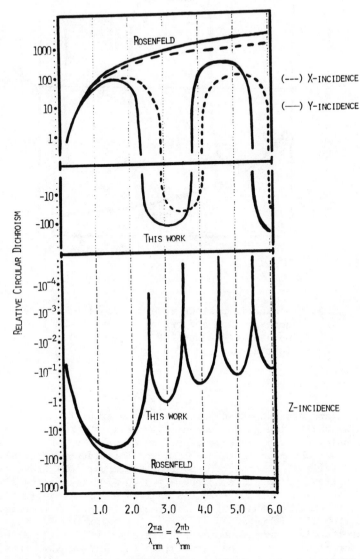

Fig. 4. A comparison of CD results from the Rosenfeld approximation and transition integrals. The relative CD is shown for the longest wavelength transition ($\Delta QN = 1$) as a function of changes in helical dimensions. The transition wavelength, λ_{nm}, was kept approximately constant as $2\pi a = 2\pi b$ were varied. The helices were right-handed and of one-turn.

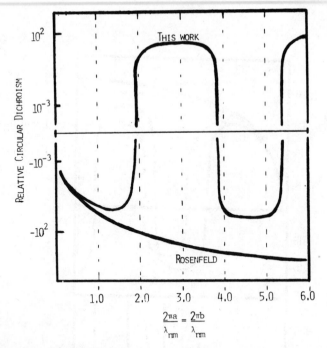

Fig. 5.   A comparison of CD results from the Rosenfeld
          approximation and transition integrals for Y
          incident light and ΔQN = 2. The helices were
          right-handed and of one-turn.

lical dimensions are increased relative to the transition wavelength, the
Rosenfeld approximation gives increasingly negative CD magnitudes, whereas the transition integrals give a pattern of negative and positive CD
similar to that observed for ΔQN = 1, X and Y incidence (Fig. 4). For
X, Y incident light, all shorter wavelength (larger ΔQN) transitions show
a similar pattern of positive and negative CD with increasing dimensions.
For Z incident light and ΔQN = 2, the pattern of CD change (not shown)
is similar to that for the longest wavelength transition (bottom, Fig.
4). Thus, for all right-handed electron on a helix models, the sign of
the CD for the first two groups of transitions (ΔQN = 1 and 2) will be
negative for light propagating parallel to the helix axis. Thus, as discussed above, the CD band signs for these two longest wavelength bands
should be useful in determining the handedness of electron on a helix
type systems of unknown dimensions. For shorter wavelength transitions,
the CD for Z incident light may be positive or negative depending upon
the value of $2\pi b/\lambda_{nm}$. For example, shown in Fig. 6 are CD results (for
Z incident light, ΔQN = 3) as $2\pi b/\lambda_{nm}$ is increased for right-handed, one-

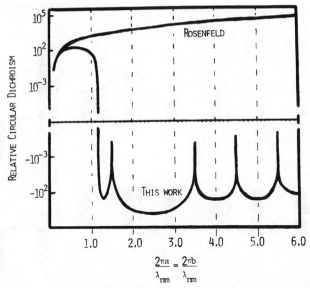

Fig. 6. A comparison of CD results from the Rosenfeld
          approximation and transition integrals for Z
          incident light and ΔQN = 3. The helices were
          right-handed and of one-turn.

turn helices. When the Rosenfeld approximation is valid ($2\pi b/\lambda_{nm}$ < 1,
Fig. 6) the CD is positive. As the dimensions increase, the transition
integral CD eventually becomes negative and remains so with further increases in dimensions. For even higher values of ΔQN, similar CD patterns
are observed.

From the above, it appears that the most interesting feature in
studying oriented systems of the electron on a helix type is the CD for
Z incident light. The results shown in Fig. 7 illustrate this. For Z
incident light, the CD for the first nine longest wavelength transition
groups (ΔQN = 1 to 9, from right to left in Fig. 7) was calculated for
three different right-handed, one-turn helical systems. Results from the
Rosenfeld approximation and transition integrals are compared. For the
first helical system (top, Fig. 7) the helical dimensions were chosen so
that the Rosenfeld approximation would be valid for all nine transition
groups. The transition integral CD (solid bars, Fig. 7) is observed to
be allowed for all transitions whereas the Rosenfeld CD is allowed only
for ΔQN = odd values and for ΔQN = 2k (Table 1). The magnitudes of the
allowed Rosenfeld CD bands are observed to be identical to those from the
transition integrals (top, Fig. 7). The first two longest wavelength CD
bands (ΔQN = 1 and 2, top, Fig. 7) are both negative while all shorter
wavelength CD bands are positive. For the second helical system (center,

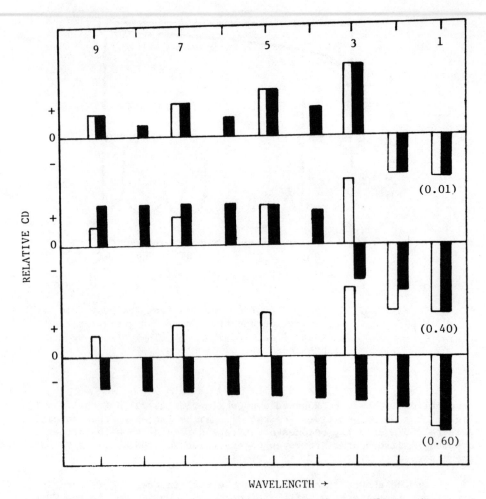

Fig. 7   The relative CD for Z incident light for the first nine
         longest wavelength bands of three helical systems. The
         numbers, 1 - 9, indicate the value of ΔQN. Solid bars
         and open bars indicate the CD band magnitudes from the
         transition integrals and Rosenfeld approximation, re-
         spectively. The value of $2\pi b/\lambda_{nm}$ for the longest wave-
         length transitions is shown in parentheses. The value
         of $2\pi b/\lambda_{nm}$ for shorter wavelength bands can be found
         by multiplying the value of the longest wavelength band
         by ΔQN for the band in question. The wavelength spac-
         ings between equivalent bands of the the three systems
         are the same, however, the spacings between adjacent
         bands are not all the same, contrary to the above.

Fig. 7), the helical dimensions have been increased substantially over
those used for the results given in the top of Fig. 7. At the helical
dimensions used (center, Fig. 7) the Rosenfeld approximation is beginning
to breakdown for most transitions. For example, there are now three negative longer wavelength bands calculated from the transition integrals,
whereas the Rosenfeld approximation predicts only two. For the third he-

lical system (bottom, Fig. 7) of even larger dimensions, the Rosenfeld approximation is valid only in terms of the signs of the first two longer wavelength bands. For $\Delta QN = 1 - 9$, the transition integrals give only negative bands. Eventually, at shorter wavelengths than shown, the longer wavelength series of negative bands is followed by positive bands. Thus, for Z incident light of a right-handed electron on a helix system, one will always find a series of negative CD bands at longer wavelengths followed by positive CD bands at shorter wavelengths.

There are real chiral systems with optical properties that follow the electron on a helix model, as has been demonstrated by Moore and Tinoco (1980). Other chiral systems that may have similar optical properties are biological helices found in viruses, chromosomes, and cells (Dorman and Maestre, 1973; Sipski and Wagner, 1977; Cowman and Fasman, 1978). The origins of the CD bands of such large molecular aggregates are not well understood. It is known that the dimensions of these chiral systems are large relative to the wavelength of light giving rise to their optical properties. Orienting such systems by physical means is not difficult, thus the applicability of the electron on a helix model in terms of helical excitons could be tested. However, since these large biological systems generally have their longer wavelength electronic transitions in the ultraviolet region of the spectrum, it would be of interest, in establishing an electron on a helix connection, to be able to probe their optical activity properties at shorter wavelengths than are now available on current instrumentation.

REFERENCES

Balazs, N. L., Brocki, T. R., and Tobias, I., 1976, Chem. Phys.,
    13:141.
Cowman, M. K. and Fasman, G. D., 1978, Proc. Natl. Acad. Sci. U.S.A.,
    75:4759.
Dorman, B. P. and Maestre, M. F., 1973, Proc. Natl. Acad. Sci.
    U.S.A., 70:225.
Moore, D. S. and Tinoco, I., 1980, J. Chem. Phys., 72:3396.
Tinoco, I. and Woody, R. W., 1964, J. Chem. Phys., 40:160.
Sipski, M. L. and Wagner, T. E., 1977, Biopolymers, 16:573.
Rosenfeld, L., 1928, Z. Phys., 52:161.

VIBRATIONAL OPTICAL ACTIVITY

Prasad L. Polavarapu

Department of Chemistry
Vanderbilt University
Nashville, TN 37235 USA

INTRODUCTION

While the measurements of optical activity in electronic transitions
are widely known and routine, similar measurements pertaining to
vibrational transitions became feasible only recently. The vibrational
optical activity (VOA) measurements can now be carried out with greater
confidence owing to the rapid developments in both instrumentation and
theory. The emergence of VOA, as a combination of two widely practiced
branches of science namely vibrational spectroscopy and optical activity,
offered new pathways for understanding the molecular stereochemistry.
Despite its very weak nature, VOA is believed to surpass the conventional
electronic optical activity (EOA) in both informational content and
complexity. This is because in EOA studies one has to depend upon a
limited number of accessible electronic transitions, whereas in VOA
studies all 3N-6 vibrational transitions, where N is the number of atoms,
of a chiral molecule are available for probing the molecular structure.
This increased number of transitions also increases the complexity in
interpreting the VOA spectra, but one hopes to find selectivity in
structural determination. Since different vibrations encompass different
portions of a molecule, the three dimensional view at a particular
portion of the molecule may be derived from the VOA associated with the
vibrations encompassing that portion. In this way one can hope to
selectively determine the stereochemistry and assemble this information
for determining the three dimensional structure of the entire molecule.

Currently there are two avenues for measuring VOA. In one, infrared
vibrational absorption spectroscopy is employed and the difference in the
absorption for left versus right circularly polarized incident light is
measured. In the second, vibrational Raman spectroscopy is employed and
the difference in the vibrational Raman intensities for left versus right
circularly polarized monochromatic incident light is measured. Although
these two approaches, where the former is referred to as vibrational
circular dichroism (VCD) and the latter as Raman optical activity (ROA),
deal with optical activity in the same vibrational transitions, the
fundamental processes governing VCD and ROA are different. Hence,
distinctly independent stereochemical information is expected from these
two approaches.

Since the status reports on VCD and ROA are now abundant,[1-7] no
attempt is made to review the literature work here. Instead, some of the
recent VOA research that is being carried out in our own laboratory is
discussed briefly in this article.

VIBRATIONAL CIRCULAR DICHROISM

The VCD experiments in our laboratory are carried out with a Fourier
transform infrared (FTIR) spectrometer as described below. The light
emerging from a Michelson interferometer is passed through a linear
polarizer and a photoelastic modulator (PEM) operating at 95 KHz
oscillating frequency. With the orientation of polarizer's axis at 45°
to the optical axes of the PEM and an appropriate voltage setting on PEM,
alternating left and right circularly polarized states are obtained for
the emerging light components. After passing through the sample these
light components are detected by either a HgCdTe detector or a InSb
detector. The intensity modulations of light by the moving mirror in
Michelson interferometer and by the optically active sample, due to
differential absorption of left versus right circularly polarized light,
impart different alternating signals on the detector. The total signal
at the detector as a function of the mirror movement, X, in the
interferometer can be given[7] by the expressions

$$I(X) = I_{DC} + I_1 + I_2 + I_3 \tag{1}$$

where

$$I_{DC} = \int_0^{\bar{\nu}_{max}} [I_s^0(\bar{\nu}_i)/4][e^{-\alpha_R} + e^{-\alpha_L}]\, d\bar{\nu}, \tag{2}$$

$$I_1 = \int_0^{\bar{\nu}_{max}} [I_s^0(\bar{\nu}_i)/4][e^{-\alpha_R} + e^{-\alpha_L}][\cos 2\pi X \bar{\nu}_i \, d\bar{\nu} , \qquad (3)$$

$$I_2 = \int_0^{\bar{\nu}_{max}} [I_s^0(\bar{\nu}_i)/4][e^{-\alpha_R} - e^{-\alpha_L}]$$

$$[\sum_{n=1}^{\infty} 2J_{2n-1}(\delta_{\bar{\nu}_i}^0) \, \sin\{[2n-1]2\pi\omega_m t\}] \, d\bar{\nu} , \qquad (4)$$

and

$$I_3 = \int_0^{\bar{\nu}_{max}} [I_s^0(\bar{\nu}_i)/4][e^{-\alpha_R} - e^{-\alpha_L}]$$

$$[\sum_{n=1}^{\infty} 2J_{2n-1}(\delta_{\bar{\nu}_i}^0) \, \sin\{[2n-1]2\pi\omega_m t\}\cos 2\pi X \bar{\nu}_i] \, d\bar{\nu} . \qquad (5)$$

where $I_s^0(\bar{\nu}_i)$ is the intensity of the source for wavenumber $\bar{\nu}_i$, $\alpha_L = 2.303$ $A_L(\bar{\nu}_i)$, $\alpha_R = 2.303 \, A_R(\bar{\nu}_i)$ with $A_L(\bar{\nu}_i)$ and $A_R(\bar{\nu}_i)$ representing the absorbances for left and right circularly polarized light component of wavenumber $\bar{\nu}_i$, ω_m is the oscillating frequency of PEM and $J_{2n-1}(\delta_{\bar{\nu}_i}^0)$ is the Bessel function value at phase shift $\delta_{\bar{\nu}_i}^0$. If the detector signal is passed through the electronic filters which transmit only the signals at interferometer frequencies $X\bar{\nu}_i$, one would obtain a normal transmission interferogram given as

$$I_t(X) = \int_0^{\bar{\nu}_{max}} [I_s^0(\bar{\nu}_i)/4][e^{-\alpha_R} + e^{-\alpha_L}][\cos 2\pi X \bar{\nu}_i] G_f(\bar{\nu}_i) \, d\bar{\nu} . \qquad (6)$$

Similarly if the detector signal is passed through a lock-in amplifier tuned to ω_m with minimal time constant, the output of the lock-in amplifier would contain a interferogram signal superimposed on a unwanted DC signal. This DC signal can be eliminated by feeding the output of the lock-in amplifier into the same electronic filters that are used to obtain $I_t(X)$. The resulting signal, referred to as ω_m interferogram, is given as

$$I_{\omega_m}(X) = \int_0^{\bar{\nu}_{max}} [I_s^0(\bar{\nu}_i)/4][e^{-\alpha_R} - e^{-\alpha_L}]$$

$$[\cos 2\pi X \bar{\nu}_i][2J_1(\delta_{\bar{\nu}_i}^0) \, G_\ell(\bar{\nu}_i) \, G_f(\bar{\nu}_i)] \, d\bar{\nu} . \qquad (7)$$

In the previous two equations $G_f(\bar{\nu}_i)$ and $G_\ell(\bar{\nu}_i)$ represent the frequency dependent gain factors introduced by the electronic filters and lock-in amplifier respectively. The circular dichroism signal, $\Delta A(\bar{\nu}_i) = A_L(\bar{\nu}_i) - A_R(\bar{\nu}_i)$, is derived from the Fourier transforms of Eqs. (6) and (7) as

$$\frac{\int_0^{X_1} I_{\omega_m}(X) \cos 2\pi X \bar{\nu}_i \, dX}{\int_0^{X_1} I_t(X) \cos 2\pi X \bar{\nu}_i \, dX} = 2J_1(\delta \frac{0}{\bar{\nu}_i}) \ G_\ell(\bar{\nu}_i) \ \tanh\{1.15 \ \Delta A(\bar{\nu}_i)\}$$

$$\simeq 2J_1(\delta \frac{0}{\bar{\nu}_i}) \ G_\ell(\bar{\nu}_i) \ [1.15 \ \Delta A(\bar{\nu}_i)] \ . \tag{8}$$

The value of $2J_1(\delta \frac{0}{\bar{\nu}_i}) \ G_\ell(\bar{\nu}_i)$ depends on the instrumental conditions and can be obtained from a calibration curve.

Studies on Carbohydrates

One of the major anticipations from VOA spectroscopy is to be able to derive new pathways for determining the molecular structure. Since a theoretical VOA analysis inherets the usual complexities involved in vibrational analyses of polyatomic molecules, the empirical correlations relating the VOA spectra and molecular structure, analogous to the well known sector rules in EOA, are vigorously sought. From this view point, simple carbohydrates are particularly useful because they contain basically the same atomic groupings but in different geometric dispositions. In the past three years we have therefore undertaken VCD spectral studies on a series of simple carbohydrates in 1600-900 cm^{-1} region.

The most significant observation found in these studies[8] is that a VCD band located at about 1150 cm^{-1} is noticed to have a strong correlation to the sequential arrangement of hydroxyl groups. The spectrum of lyxopyranose is shown in Fig. 1, where this band can be identified. For the D-enantiomer of arabinose and lyxose the sign of the aforementioned VCD band was found to be positive, where as for that of xylose, glucose, fucose and galactose the sign of the aforementioned VCD band is found to be negative.[8] This difference can be understood from the following analysis. First consider the structure of a given sugar molecule to be composed of different chiral segments. The skeleton containing C-1 and C-2 and the hydroxyl groups associated with these atoms i.e. HO—C-1—C-2—OH represents the first segment. Viewing along the C-1—C-2 bond, if the hydroxyl group on carbon atom C-1 makes a positive dihedral angle with the hydroxyl group on carbon atom C-2, then a right handed helical contribution is imparted for this segment. For a negative dihedral angle, a left handed helical contribution is made. If the two hydroxyl groups are in axial positions, then the segment is not

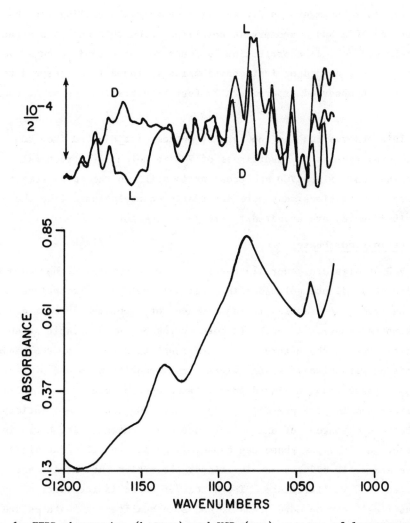

Fig. 1. FTIR absorption (bottom) and VCD (top) spectra of lyxopyranose
in pyridine-d_5 solvent. The letters D and L on the top traces
identify the VCD of appropriate enantiomer. The samples were
allowed to equilibrate in D_2O, then lypholized and the dried
samples were dissolved in pyridine-d_5. The VCD spectrum of
the racemic mixture (prepared from equal amounts of enan-
tiomers) is subtracted from that of each enantiomer to
eliminate the artifacts.

chiral and hence no contribution is made. Similar operations are
considered for the remaining segments in the ring and then the net effect
from all these segments is derived with appropriate scaling for the
percentage of α and β anomers in solution. From this analysis a net
right handed helical contribution is found to correspond to negative VCD
and a net left handed helical contribution is found to correspond to
positive VCD observed experimentally for the vibrational band at about
1150 cm^{-1}.

This observation[8] is the type of outcome anticipated from VCD
studies and reveals the usefulness of VCD in molecular structural
determination. Since VCD spectroscopy is still in the early stages of
chemical applications many more interesting correlations, like the one
described above, are expected to emerge in the future.

Studies in Vapor Phase

VCD studies are generally carried out with samples either as neat
liquids or as dilute solutions in a suitable solvent. Exceptions to this
are the studies on samples in vapor phase[9] or deposited in matrices at
low temperatures.[10] In each of these approaches some notable advantages
persist. Due to the nature of rovibrational band contours, VCD bands
associated with closely spaced vibrational transitions would overlap more
in vapor phase than in liquid phase. When such overlap does not lead to
complications in interpretation, vapor phase VCD studies are anticipated
to become the sources of unique informational content. The Coriolis
interactions in vapor phase can cause mixing of two different vibrational
states and their effects on vibrational absorption intensities are
reported[11,12] in literature. The effect of Coriolis interactions on VCD
intensities[13] can be more dramatic and some qualitative features are
described here.

The VCD intensity for a vibrational transition ℓ, represented by the
normal coordinate Q_ℓ, is determined by the product $(\partial\mu_\alpha/\partial Q_\ell)(\partial m_\alpha/\partial Q_\ell)$
where μ_α and m_α are, respectively, the electric dipole and magnetic
dipole moments. Due to the Coriolis interaction between a vibrational
motion represented by Q_r and a molecular rotation, a different
vibrational motion Q_s can be generated. In other words, a molecular
rotation can mix the two vibrational wavefunctions ψ_r and ψ_s. The new
wavefunctions may be written as linear combinations of the unperturbed
wavefunctions as

$$\psi_1 = a\psi_r + \sigma b\psi_s$$

$$\psi_2 = \sigma b\psi_r + a\psi_s \qquad (9)$$

where the mixing coefficients a and b are dependent upon the rotational quantum numbers; σ has a value of ± 1, with the parity depending[11] upon the signs of the Coriolis constants. Using the perturbed wavefunctions of the form given by Eq. (9), the VCD intensities R_1, R_2 at the band centers of the two vibrational bands participating in Coriolis interaction can be given as

$$R_1 \simeq a^2\left(\frac{\partial\mu_\alpha}{\partial Q_r}\right)\left(\frac{\partial m_\alpha}{\partial Q_r}\right) + b^2\left(\frac{\partial\mu_\alpha}{\partial Q_s}\right)\left(\frac{\partial m_\alpha}{\partial Q_s}\right)$$

$$+ \sigma a b \left[\left(\frac{\partial\mu_\alpha}{\partial Q_r}\right)\left(\frac{\partial m_\alpha}{\partial Q_s}\right) + \left(\frac{\partial\mu_\alpha}{\partial Q_s}\right)\left(\frac{\partial m_\alpha}{\partial Q_r}\right)\right] \qquad (10)$$

$$R_2 \simeq a^2\left(\frac{\partial\mu_\alpha}{\partial Q_s}\right)\left(\frac{\partial m_\alpha}{\partial Q_s}\right) + b^2\left(\frac{\partial\mu_\alpha}{\partial Q_r}\right)\left(\frac{\partial m_\alpha}{\partial Q_r}\right)$$

$$+ \sigma a b \left[\left(\frac{\partial\mu_\alpha}{\partial Q_r}\right)\left(\frac{\partial m_\alpha}{\partial Q_s}\right) + \left(\frac{\partial\mu_\alpha}{\partial Q_s}\right)\left(\frac{\partial m_\alpha}{\partial Q_r}\right)\right] \qquad (11)$$

Different permutations of the signs of electric and magnetic dipole derivatives for the interacting modes are possible as shown in Table 1.

Table 1.   Signs of Dipole Derivatives for Vibrational Modes
Participating in Coriolis Interactions.

$\left(\frac{\partial\mu_\alpha}{\partial Q_r}\right)\left(\frac{\partial m_\alpha}{\partial Q_r}\right)$	$\left(\frac{\partial\mu_\alpha}{\partial Q_s}\right)\left(\frac{\partial m_\alpha}{\partial Q_s}\right)$	$\left(\frac{\partial\mu_\alpha}{\partial Q_r}\right)\left(\frac{\partial m_\alpha}{\partial Q_s}\right)$	$\left(\frac{\partial\mu_\alpha}{\partial Q_s}\right)\left(\frac{\partial m_\alpha}{\partial Q_r}\right)$
(-)(-)=+	(-)(-)=+	+	+
(-)(-)=+	(+)(+)=+	-	-
(+)(+)=+	(-)(-)=+	-	-
(+)(+)=+	(+)(+)=+	+	+
(-)(+)=-	(-)(+)=-	-	-
(-)(+)=-	(+)(-)=-	+	+
(+)(-)=-	(-)(+)=-	+	+
(+)(-)=-	(+)(-)=-	-	-
(-)(-)=+	(-)(+)=-	-	+
(-)(-)=+	(+)(-)=-	+	-
(+)(+)=+	(-)(+)=-	+	-
(+)(+)=+	(+)(-)=-	-	+
(-)(+)=-	(-)(-)=+	+	-
(-)(+)=-	(+)(+)=+	-	+
(+)(-)=-	(-)(-)=+	-	+
(+)(-)=-	(+)(+)=+	+	-

Now let us consider two interacting bands one with positive VCD and
another with negative VCD of equal magnitude. If the magnitudes of cross
products $(\partial\mu_\alpha/\partial Q_r)(\partial m_\alpha/\partial Q_s)$ and $(\partial\mu_\alpha/\partial Q_s)(\partial m_\alpha/\partial Q_r)$ are equal, and the two
vibrational wavefunctions are strongly mixed (i.e. $a\simeq b\simeq 1/\sqrt{2}$) at lower
rotational quantum numbers, then the VCD intensity of both bands will be
depleted at the band center. This consideration is more suited for the
vibrations of a methyl group because, the degenerate vibrations of a
methyl group, which will be split in a chiral environment, can have
further splitting due to the usually strong Coriolis interaction among
them. These split modes can have bisignate VCD of equal magnitude as
well as the limiting values of $1/\sqrt{2}$ for the mixing coefficients a and b
at lower rotational quantum numbers.

    Another interesting situation arises when one mode, say Q_s, has a
very small electric dipole transition moment, and the magnitudes of
magnetic dipole transition moments for Q_r and Q_s are nearly equal. Then
only one cross term, $(\partial\mu_\alpha/\partial Q_r)(\partial m_\alpha/\partial Q_s)$, in Eqs. (10) and (11) survives.
In this situation, the analysis of intensity distribution in the wings of
these bands can be carried out, as done by Mills[11,12] for normal
absorption spectra, and the relative signs of the derivatives $(\partial\mu_\alpha/\partial Q_r)$
and $(\partial m_\alpha/\partial Q_s)$ may be derived with some approximations. However, only
gross features of Coriolis effects on VCD could be observed[9] so far.
Detailed effects are not yet observed due to the difficulties involved in
measuring high resolution VCD spectra. At present a VCD measurment in
$1600-800$ cm^{-1} region at 4 cm^{-1} resolution requires at least two hours of
data acquistion time for each enantiomer on a FTIR-VCD spectrometer. To
measure VCD with higher resolution a longer mirror stroke in the
interferometer of an FTIR spectrometer would be required and this can
introduce severe noise into VCD measurements (even for 1 cm^{-1}
resolution). Hence it appears unlikely that the detailed Coriolis
effects on VCD can be observed experimentally at the present time. But
gross effects, like decreased VCD intensity, can be observed at low
resolution.

RAMAN OPTICAL ACTIVITY MEASUREMENTS

    The measurements of Raman optical activity in the early
developmental stages were somewhat uncertain due to the artifacts in ROA
measurements. The nature of these artifacts and the procedures to
eliminate them are now better understood.[2,14] The early ROA spectra were

measured on scanning Raman spectrometers with a photomultiplier tube
detection. This procedure turned out to be notoriously tedious because
the higher photon counts required for measuring small ROA magnitudes
necessiated slow scan rates, typically 1 cm^{-1} per minute. Despite this
disadvantage, Barron[2] succeeded in recording abundant ROA spectra and
kept ROA spectroscopy not only alive but also at par with its counter
part, namely VCD spectroscopy.

Recent developments in multichannel detector technology provided the
opportunities for faster ROA spectral acquisition. Moskovits,[15] Hug[14]
and their coworkers have successfully demonstrated the advantages of
multichannel detectors for ROA spectroscopy. For a better perspective of
these developments, and also for basic theory of ROA, and numerous
experimental ROA spectra the reviews of Barron[2] should be consulted.

At Vanderbilt, we have undertaken the plans for a multichannel ROA
instrument following the earlier ROA measurements[16] with a scanning Raman
spectrometer at Syracuse. The design of our multichannel ROA
spectrograph has been described elsewhere.[17] Here we present the details
of electronic signals used in ROA data collection, along with
representative ROA spectra to indicate the performance of our instrument.

The basic instrument contains a Ar$^+$ laser (Spectra physics, Model
165-06), collection optics, 0.5 meter spectrograph (Spex, Model 1870)
with 1200 gr/mm grating and a intensified diode array detector (Tracor
Northern, Model TN-1223-4I) with 1024 detector elements on 25 m centers.
Slight modifications incorporated into our earlier[17] collection optics
are as follows: (a) For a better alignment of the sample in the laser
beam, a fiber optic coupler mount (Newport Research) is used to hold a
lens (focusing the beam to 5.8 m size) on one end and the sample on the
other end; (b) the collecting lenses are replaced by Nikon F/1.8E lenses
which provided extra freedom in focusing the scattered beam.

The detector is controlled by a multichannel analyzer (Tracor
Northern, TN-1710) but for ROA data collection some significant changes
or additional interfacing circuitry were found necessary. Three
different facilities available on the analyzer are used to derive a
reliable ROA data collection. A TTL square wave from a function
generator or a microsecond pulse sequence from a dedicated microcomputer
(Radio Shack, Model IV) at 2 Hz frequency is fed into BNC A on the
analyzer. When a falling edge of this TTL signal is sensed by the

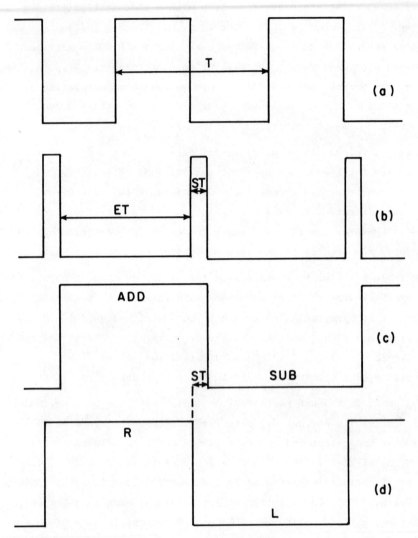

Fig. 2. Electronic signals used for ROA data acquisition. When a TTL
square wave (a) with T secs for each cycle is supplied, at each
falling edge of this signal a TTL pulse for scan duration
(ST) is provided by the analyzer (TN-1710). ET is the expo-
sure time for the detector elements. Signals (c) and (d) are
synchronous to and generated from (b). When signal (c) is
HIGH the digitization data is stored in ADD mode and when LOW
the data is stored in SUB mode. Signal (d) is used to drive
an electrooptic modulator that imparts right (R) and left (L)
circularly polarized states for the incident laser light.

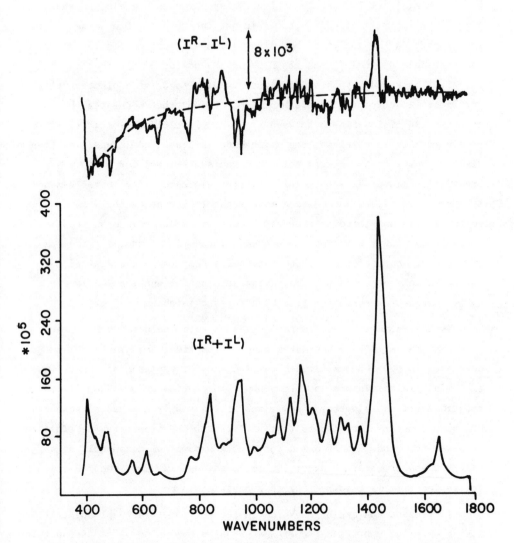

Fig. 3. Depolarized Raman (bottom) and ROA (top) spectra for (+)-α-
pinene. The data collection time is approximately 2.5 hours
with 500 mw laser power at 488 nm. ROA spectrum has a sloping
background as represented by the dashed line.

analyzer, a detector scan is performed. The scan duration is
approximately 10 msec (10 microseconds per detector element) and for this
duration a synchronous TTL pulse is provided by the analyzer at GATE OUT
port. This signal is used to derive two synchronous signals, as shown in
Fig. 2, by a simple circuitry built in our laboratory. One signal is fed
into a port on the analyzer that directs the data collection in add or
subtract mode depending upon the HIGH or LOW status of the incoming
signal. The second synchronous signal is used to drive a electrooptic
modulator (Electrooptic Development Ltd., PC105) that modulates the
polarization of laser beam between right and left circular states. This
scheme results in obtaining the difference in scattered intensities (for
right versus left circularly polarized incident light) for Raman
frequencies dispersed across 1024 detector elements. The normal Raman
spectrum is obtained in additive mode separately. After the data
collection is completed, the spectral data is transferred from the
analyzer to floppy discs of the aforementioned microcomputer through a
RS232C port and therefrom to the University Computer, when desired,
through a telephone modem. To indicate the performance, a ROA spectrum
of (+)-α-pinene recorded on this spectrograph is shown in Fig. 3.

Typical conditions for our ROA measurements include 500 mW laser
power at 488 nm, 2 to 3 hour data acquisition time, and simultaneous
detection in 400–1800 cm^{-1} region. Hence an ROA spectrum of about 1400
cm^{-1} width is obtainable in 2 to 3 hour time period on our ROA
spectrometer. This performance is superior not only to that of scanning
ROA spectrometers but also to that of FTIR–VCD spectrometers.[6,7] This is
because, the bandwidth of a mid infrared VCD spectrum measurable on a
FTIR–VCD spectrometer is about 800 cm^{-1}, and usually 2 1/2 hour data
acquistion time is required for this bandwidth.

At present ROA can be measured in the entire vibrational spectral
region where as VCD has not yet been measured below 600 cm^{-1}. Below this
frequency range, VCD measurements await the development of suitable
photoelastic modulators. For this reason and also since ROA, in
comparison to VCD, can be measured in a larger vibrational bandwidth
(with a 600 gr/mm grating it would be possible to record, on our
spectrograph, 100–2000 cm^{-1} region) at once, the appeal for ROA
spectroscopy over VCD spectroscopy is continuously increasing.

ACKNOWLEDGEMENT

Several communications with Professor L. D. Barron, sharing his
experiences with his multichannel ROA instrument are gratefully
acknowledged. This work is supported by the grants from NIH (GM 29375),
Research Corporation and Vanderbilt University. Acknowledgement is made
to the Donors of Petroleum Research Fund Administered by the American
Chemical Society for partial support. Technical assistance from
Mr. S. Bottoms in the assembly of some electronic circuitry is also
acknowledged.

REFERENCES

1. P. J. Stephens, and R. Clark, Vibrational circular dichroism: The
    experimental view point, in: "Optical Activity and Chiral
    Discrimination", S. F. Mason, ed., D. Reidel, Dordrecht (1979).

2. L. D. Barron, and J. Vrbancich, Natural vibrational Raman optical
    activity, Top. Curr. Chem., 123:151 (1984).

3. L. A. Nafie, Infrared and Raman vibrational optical activity,
    in: "Vibrational Spectra and Structure", J. R. Durig, ed., Vol. 10,
    Elsevier, Amsterdam (1981).

4. T. A. Keiderling, Vibrational circular dichroism, Appl. Spectrosc.
    Rev., 17:189 (1981).

5. P. L. Polavarapu, Recent advances in model calculations of
    vibrational optical activity, in: "Vibrational Spectra and
    Structure", J. R. Durig, ed., Vol. 13, Elsevier, Amsterdam (1984).

6. L. A. Nafie and D. W. Vidrine, Double modulation Fourier transform
    spectroscopy, in: "Fourier Transform Infrared Spectroscopy", J. R.
    Ferraro and L. J. Basile, eds, Vol. 3, Academic Press, New York
    (1982).

7. P. L. Polavarapu, Fourier transform infrared vibrational circular
    dichroism, in: "Fourier Transform Infrared Spectroscopy", J. R.
    Ferraro and L. J. Basile, eds., Vol. 4, Academic Press, New York (in
    press).

8. D. M. Back and P. L. Polavarapu, Fourier transform infrared
    vibrational circular dichroism of sugars: A spectra-structure
    correlation, Carbohyd. Res., (in press).

9. P. L. Polavarapu and D. F. Michalska, Vibrational circular dichroism in (S)-(-)-epoxypropane; Measurement in vapor phase and verification of the perturbed degenerate mode theory, J. Am. Chem. Soc., 105:6190 (1983).

10. D. W. Schlosser, F. Devlin, K. Jalkanen and P. J. Stephens, Vibrational circular dichroism of matrix isolated molecules, Chem. Phys. Lett., 88:286 (1982).

11. I. M. Mills, Coriolis interactions intensity perturbations and potential functions in polyatomic molecules, Pure and Appl. Chem., 11:325 (1965).

12. C. DiLauro and I. M. Mills, Coriolis interactions about x-y axes in symmetric tops, J. Mol. Spectrosc., 21:386 (1966).

13. P. L. Polavarapu, Vibrational circular dichroism in liquid and vapor phase, Bull. Am. Phys. Soc., 28:1343 (1983).

14. W. Hug, Optical artefacts and their control in Raman circular difference scattering measurements, Appl. Spectrosc., 35:115 (1981).

15. T. Brocki, M. Moskovits and B. Bosnich, Vibrational optical activity: Circular differential Raman scattering from a series of chiral terpenes, J. Am. Chem. Soc., 102:495 (1980).

16. P. L. Polavarapu, M. Diem and L. A. Nafie, Vibrational optical activity in para-substituted 1-methylcylohex-1-enes, J. Am. Chem. Soc., 102:5449 (1980).

17. P. L. Polavarapu, A design of Raman spectrograph for optical activity and normal Raman measurements, Appl. Spectrosc., 37:447 (1983).

VACUUM ULTRAVIOLET CIRCULAR DICHROISM STUDIES OF

PEPTIDES AND SACCHARIDES

Eugene S. Stevens

Department of Chemistry
State University of New York
Binghamton, New York 13901

INTRODUCTION

Conventional circular dichroism (CD) spectrometers operate in the ultraviolet region to approximately 185 nm. Peptides and proteins have only two optically active backbone transitions in this region, the n-π^* and the π-π^* amide transitions, and even then the entire π-π^* transition is not generally observable with commercial spectrometers since its envelope can extend to 175–185 nm. Vacuum ultraviolet circular dichroism (VUCD) measurements are required to observe the high energy component of the π-π^* transition and the two higher energy transitions near 165 nm and 145 nm, all of which have been shown to be sensitive to molecular conformation.

Vacuum ultraviolet spectroscopy is absolutely required for saccharides because unsubstituted saccharides have no electronic transitions above 190 nm. Without vacuum ultraviolet measurements, the optical activity of unsubstituted saccharides is displayed only through the near ultraviolet optical rotatory dispersion (ORD), which displays no Cotton effects or extrema. Thus, whereas CD studies have helped reveal the rich conformational behavior of proteins and nucleic acids, saccharides have remained out of the range of such studies until recently because of technical limitations inherent in conventional CD spectrometers. Furthermore, the entire area of theoretical description of higher energy electronic states of peptides and saccharides is enlarged with the recently developed field of VUCD spectroscopy.

173

Prototype VUCD instruments of current design were first described by Schnepp,[1] Johnson,[2] Pysh(Stevens),[3] Brahms,[4] and Bush.[5] All of these instruments use a hydrogen discharge as light source. Synchrotron radiation is an intrinsically stronger light source since its intensity increases with beam current. It therefore provides the possibility of measuring VUCD with a sensitivity heretofore not possible.[6] Sutherland[7] and Snyder[8] have demonstrated the application of synchrotron radiation to VUCD measurements.

In all VUCD measurements to date the quarter-wave retarder, which converts plane polarized radiation to circularly polarized radiation, is a photoelastic modulator constructed of material (CaF_2) having a transmission limit of approximately 130 nm. Synchrotron radiation, besides its potentially great intensity in the vacuum ultraviolet region, allows the possibility (described elsewhere in this monograph) of being circularly polarized without a transmission device which, if realized, would remove that limitation.

The purpose of this article is to describe VUCD measurements on peptides and saccharides carried out in our laboratory over the past several years. Recent examples of VUCD measurements on related systems are found in the works of Johnson,[9-19] Brahms,[20-26] Bush,[27-33] Sutherland,[34-36] and Snyder.[8]

INSTRUMENT DESIGN

Our prototype VUCD spectrometer[3] is similar in design to the others mentioned above, with relatively minor differences. The monochromator is a McPherson vacuum instrument with a slit-to-grating distance of 1 meter and first order reciprocal linear dispersion of 1.66 nm/mm. The grating has 600 lines/mm and a MgF_2,Al coating for high reflectivity.

The light source is a cold cathode hydrogen discharge lamp of the Hinteregger type which provides a broad continuum above 165 nm; below 165 nm is the relatively more intense many-lined molecular hydrogen spectrum. The discharge is contained in a 3 mm ID quartz capillary with long axis directed toward the slit so that the discharge is viewed "end on" for maximum intensity. A 1 mm thick CaF_2 window iso-lates the lamp from the monochromator so that larger entrance slits,

and concommitant light levels, are possible. Hydrogen is bled through
the lamp slowly while maintaining a pressure of 100-200 N/m^2 with a
separate pumping system. Normally a discharge current of 0.5 A is used
which results in a 700 V drop across the discharge.

The polarizer is similar to the type described by Robin, Kuebler
and Pao[37] and was supplied by McPherson Corporation. It consists of a
biotite $\left[K_2(Mg,Fe,Al)_6(Si,Al)_8O_{20}(OH)_4\right]$ plate polarizer and mirror
assembly. The photoelastic modulator, from Morvue Electronic Systems,
is used in the "split-head" configuration in which the optical-transducer
element is mounted in a housing separate from the rest of the oscillator
circuitry, which permits the optical element to be housed in a moderately
sized sample chamber. The photomultiplier has a quartz window and is
used either with or without a sodium salicylate coating. The phosphor
coating extends the detection range from the quartz cutoff near 160 nm
to the instrument limit of approximately 130 nm, determined by the
photoelastic modulator. The phosphor converts incident radiation to
425 nm light. The quantum efficiency of sodium salicylate is approximately 65% from below the CaF_2 cutoff to 340 nm.[38] The fluorescent
decay time is about 10 ns,[38] which is small compared to the period of
oscillation of the modulator (20 μs). The coating density is approximately 1 mg/cm^2.

The modulation voltage of the quarter wave plate is programmed by
mechanical linkage to the monochromator wavelength drive to maintain
quarter wave retardation at all wavelengths. The DC signal at the
photomultiplier is kept constant by modulating the photomultiplier gain
in response to the lamp output; the AC signal at the photomultiplier is
proportional to the ellipticity of the sample. The AC signal is sent
to a lock-in detector amplifier, and then to a two channel strip chart
recorder. We calibrate our instrument with   d-10-camphorsulfonic acid
according to standard procedures.[39]

PEPTIDES

In our first application of VUCD to the study of biologically
relevant materials, we demonstrated that the VUCD of peptides was in
fact conformation dependent by examining the known helical structures
of polymers of alanine[3] and proline.[40] Thereupon we undertook a

comprehensive study of the β-pleated sheet conformation of aggregated peptides in collaboration with Prof. C. Toniolo of the University of Padua. Our motivation was the lack of any experimental means for distinguishing the parallel and antiparallel β-pleated sheets (see Fig. 1). A 1690 cm^{-1} IR band is accepted as a positive test for the antiparallel sheet; but the absence of that band is not acceptable as a positive test for the parallel sheet. The plausibility that VUCD could distinguish the two sheets stemmed from theoretical work of our own[41,42] and others.[43] In the parallel arrangement, theory indicates only one positive CD component for the π–π* band leading to a CD crossover from positive to negative dichroism near 190 nm; in the antiparallel arrangement the existence of two positive CD components for the π–π* band leads to a crossover at significantly lower wavelength, near 178 nm, in calculated spectra.

We found the predicted variation in CD crossover and, on the basis of our studies of numerous synthetic linear homooligopeptides, have made conformational assignments as shown in Table 1.

Table 1. β-Sheet assignments (parallel or antiparallel) of linear homooligopeptides obtained from vacuum UV CD studies.

Antiparallel (178–179 nm crossover)	Mixed Parallel and Antiparallel (184–186 nm crossover)	Parallel (190–192 nm crossover)
Alanine (Refs. 44, 45)	Norvaline (Ref. 44)	Valine (Refs. 44, 45)
L-α-amino-n-butyric acid (Ref. 46)	Leucine (Ref. 47)	Phenylalanine (Refs. 51, 52)
	Methionine (Ref. 48)	
	Norleucine (Ref. 49)	
	Methylcysteine (Ref. 50)	

We conclude that parallel chains are favored by an overall bulkiness in the side chain and by the presence of β-branching in the side chain. To rationalize this behavior we have suggested the role of unfavorable contacts between neighboring side chains in the antiparallel arrangement analogous to 1,3-axial interactions in cyclohexane derivatives. For small side chains where such contacts are not present, the general favoring of the antiparallel arrangement is not overridden. We

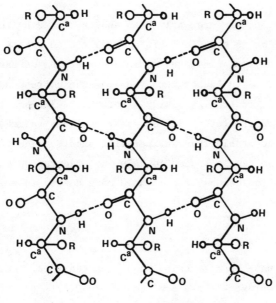

(a)   ANTI-PARALLEL *β*-SHEET

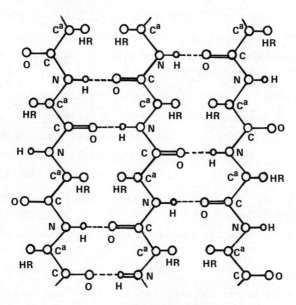

(b)   PARALLEL *β*-SHEET

Figure 1.   The antiparallel (a) and parallel (b) peptide β-sheets.

consider our work on the associated β-forms of peptides essentially
complete.

Our conclusions have recently been substantiated by the detailed
energy minimization calculations of Scheraga and coworkers.[53-55] They
found that, for poly-alanine, the minimized energies of parallel β-
sheets are considerably higher than those of the corresponding antiparallel β-sheets, indicating that the antiparallel β-sheets are
intrinsically more stable.[53] In contrast, for poly-valine, parallel
β--sheets are more stable than antiparallel sheets. They found that
valine side chains can be placed in only one orientation in β-sheets
and that orientation can be accommodated more easily in the parallel
structure.[54,55]

SACCHARIDES

Recently we have been focusing on polysaccharides. Conventional
non-vacuum CD instruments have limited applicability to the study of
polysaccharides because the electronic transitions of the polysaccharide backbone occur at wavelengths below 190 nm. Although some information about the behavior of these transitions has been obtained in-
directly from ORD measurements at higher wavelength, direct measurement
of polysaccharide backbone CD has been possible only since the development of vacuum CD instruments. The work of our own laboratory has
been on carrageenan,[56] agarose,[57] dextran,[58] galactomannans,[59]
pustulan,[60] glucans in general,[61] acetylated glucans,[62,63] hyaluronic
acid,[64] chitin,[65] alginates,[66] chondroitins,[67] and galacturonates.[68]
One long-term objective has been to delineate the contributions of the
potentially large number of factors determining VUCD spectra (e.g.,
anomeric configuration, linkage type, and intra- and intermolecular
hydrogen bonding). Comparison of film, gel, and solution spectra is
important since the film spectra can often be correlated with results
from x-ray crystallography.

The first generalization that resulted from this work was that
unsubstituted polysaccharide chains have two major CD bands in the
vacuum UV, one in the region 165-180 nm and the other in the region

145-160 nm. For a given polysaccharide these two bands are always
of opposite sign, but chemical structure and configuration appear to
determine whether the positive band is the low energy or high energy
band of the pair.

In collaboration with E. R. Morris and D. A. Rees (formerly with
Unilever Research, Colworth House, Bedford, now with the National
Institute for Medical Research, Mill Hill, London), we carried out a
detailed study of galactomannans.[59] Galactomannans are plant polysaccharides having structures based on a β(1→4) linked D-mannose backbone, with α(1→6) linked D-galactose substituents, as shown in Fig. 2.
The proportion and distribution of galactose shows substantial variation with botanical origin. We used three galactomannan samples
derived from the seed endosperm of *Certaonia siliqua*, *Caesalpinia
spinosa*, and *Cyamopsis tetragonolobus*, referred to subsequently by
their trivial names of carob, tara, and guar, respectively. Molecular
interactions of galactomannan chains, both in self-association and in
the formation of mixed junctions with other polysaccharides, involve
unsubstituted, or sparingly substituted, regions of mannan backbone.
Such interactions are most evident for carob, and decrease through
tara to guar, which is relatively heavily substituted.

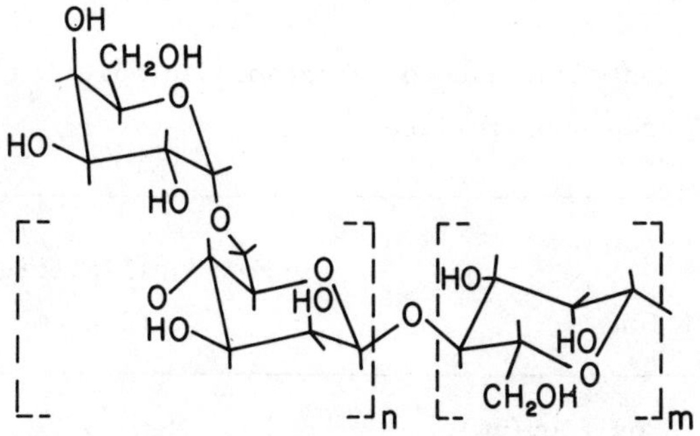

Figure 2. Galactomannan primary structure

The specific objectives of the study were to measure the composition dependence of galactomannan VUCD, to extract chiroptical parameters
for pure galactan and pure mannan polymers, and, by correlating the
solution and solid film VUCD data with results of x-ray fiber diffraction studies, to establish the contributions from the various factors
determining VUCD spectra in the solution, gel and solid states.

VUCD solid film spectra were recorded to 140 nm, and in all cases
show a positive band at 169 nm, and a negative band at 149 nm whose
relative intensity increases systematically with decreasing galactose
content. In solution only the lower energy band is accessible, and
has the same position and width as in the solid state but substantially greater amplitude. Residual ORD behavior, after subtraction
of the contribution from the 169 nm band (calculated by Kronig-
Kramers transform of fitted CD parameters) shows a single band of the
same position and width as the high energy solid state transition.

Thus, galactomannan optical activity in solution, as well as in
solid films, can be described adequately in terms of two bands of fixed
position (~169 and ~149 nm) and width (~10.8 nm in both cases). Vari-
tion in the intensities of these bands with galactomannan composition
is linear for both transitions and indicates simple additivity of
contributions from the two different monomers present. Extrapolation
to 0 and 100% galactose content yields the individual molar ellipticities for mannan and galactan listed in Table 2. For both transitions
the contributions from the two components are of opposite sign, but
the relative magnitudes are substantially different.

Table 2. Contributions to galactomannan solution VUCD from
          component residues.

Transition		Molar ellipticity, 10^{-3}x $[\theta]$ (deg cm^2 $dmol^{-1}$)	
Width(nm)	Position(nm)	Mannan	Galactan
10.8±0.5	169±1	+14	-8
10.8±0.5	149±1	-33	+80

Two CD bands of opposite sign in the vacuum UV have also been observed for carrageenan,[56] agarose,[57] amylose,[61] and alginate.[61] For all of these the higher energy band is the more intense, and dictates the sign of optical rotation at longer wavelengths, such as at the sodium D-line which is commonly used in characterization of carbohydrates. In the case of the galactomannans, by contrast, D-line rotation is the same sign as the lower energy band, and shows systematic variation with composition. The origin of this behavior is evident from Table 2. For both the mannan and galactan components the higher energy (149 nm) band is considerably more intense than the lower energy (169 nm) band, and will therefore determine the sign of the high wavelength optical rotation contribution from that component, as in the case of all other polysaccharides so far studied. The mannan and galactan contributions to both transitions, however, are opposite in sign, which has the effect of making the net intensity of the two bands more nearly equal, so that proximity to the wavelength at which optical rotation is measured becomes of greater significance than intensity.

It is now well established that polysaccharide optical rotation behavior is sensitive to conformation, as well as to primary structure. Our results indicate that galactomannan optical rotation has its origin entirely in the two VUCD backbone transitions at 149 and 169 nm. Thus, any changes in optical rotation with galactomannan conformation must arise from changes in the intensity of one or both of these bands.

X-ray fiber diffraction studies of galactomannans show an extended two-fold conformation of the mannan backbone in the solid state.[69-71] Potential energy calculations[72] suggest only limited conformational mobility about this preferred conformation, which is confirmed by viscosity measurements showing extended random-coil dimensions in solution,[73] consistent with relatively inflexible chain geometry. It is well established, however, that 1→6 linkages such as those of the galactose substituents, are extremely flexible, since the C(5)-C(6) bond introduces an additional degree of conformational freedom. It is likely, therefore, that in solution orientations of galactose relative to the polymer backbone will be essentially random. Thus, we would expect only limited contribution from the mannan chain to the dissymmetric environment of the galactose VUCD chromophores.

In the solid state the net ellipticities for both the 169 and 149 nm bands are much lower than in solution for all three galactomannans. If the constraints of solid state packing impose any conformational restriction or orientation upon the side chains, we would anticipate an enhancement of the circular dichroism contribution from these residues. Since (Table 2) their CD behavior is opposite in sign to that of the mannan backbone, this would lead to a reduction in net intensity. The fact that such a reduction does occur indicates a conformational ordering of galactose side chains on going from solution to the solid state.

Gels formed on freezing and rethawing carob solutions[74] show approximately 20% reduction in intensity of the accessible lower energy transition. After melting the gel network by heating to 80°C, the original solution CD is restored. Galactomannans of higher galactose content show no such gelation behavior. The observed decrease in the ellipticity of the 169 nm band indicates partial adoption of solid state geometry in intermolecular junctions of freeze-thaw carob gels.

Our studies of D-glucans[58,60-63] included the compounds shown in Table 3. In the region 164–177 nm all α-D-glucans show positive dichroism, suggesting that the chiroptical properties of the transition responsible for that dichroism are relatively independent of molecular conformation and linkage type for α-D-glucans.

Nelson and Johnson[9] have suggested that there is a correlation in the sign of the 164–177 nm band and the anomeric configuration of methyl pyranosides, but they observed an exception to the correlation in the case of glucopyranosides; i.e., even at the monosaccharide level, the correlation between sign of the 164–177 nm CD band and anomeric configuration is not strict. Our work shows that there is also no strict correlation in D-glucans between the sign of the CD in the 164–177 region and anomeric configuration. There is a correlation for 1→3 and 1→4 D-glucans, but for 1→6 D-glucans (dextran and pustulan) there is only a decrease in (positive) CD in the 164–177 nm region on going from the α- to β-linked chain, with no change in sign. Thus, in D-glucans the dependence of VUCD on C(1) configuration is modulated by linkage site, with the C(6) hydroxymethyl group as linkage site playing a special role. This conclusion is consistent with the previously

Table 3. D-Glucans

D-Glucan	Branched Linkages	Common Name
(1→3)-α-	–	Pseudonigeran
(1→3)-β-	–	Curdlan
(1→4)-α-	–	Amylose
(1→4)-α-	~5% (1→6)-α-	Amylopectin
(1→4)-α-	~15% (1→6)-α-	Glycogen
(1→4)-β-	–	Cellulose
(1→6)-α-	~5% (1→3)-α-	Dextran
(1→6)-β-	–	Pustulan
(1→3)-α-(1→4)-α-	–	Nigeran
(1→6)-α-[(1→4)-α-]$_2$	–	Pullulan

reported result[9] that the C(6) hydroxymethyl group also plays a significant role in determining the sign of the 164-177 nm CD of monomeric
glucopyranoses.

It was during the study of glucans that we discovered conditions
under which a low energy CD band appears in unsubstituted polysaccharide
chains above 180 nm. Thus, whereas a freshly prepared solution of
pustulan shows only positive CD below 180 nm, a negative band centered
near 190 nm begins to develop as pustulan gels. The gelation itself
was unexpected because the (1→6) linkage generally imparts sufficient
flexibility to inhibit gelation, as in the case of dextran. As pustulan
gels the CD band near 190 nm intensifies and blue shifts. Gelation of
polysaccharides generally involves the formation of gel-junction zones
in which the chains become ordered. Thus, the increase in intensity of
the negative CD band with time and its blue shift can be associated
with the development of ordered conformations and their aggregation to
form the gel.

In amylose gels a negative CD band near 182 nm is also observed
but the band is observed in solution as well. This result is in accord
with a pseudohelical conformation of amylose being present in solution which, upon gelation, is only slightly modified through the associ-
ation of helical molecules and the formation of junction zones. Amylo-

pectin, glycogen and pullulan all display weaker negative CD bands near
190 nm reflecting a relative randomization of molecular conformation
resulting from the presence of (1→6)-α linkages.

Our conclusion, therefore, is that the negative CD band we see in
the 180-190 nm region of some D-glucans reflects a high degree of local
order in the polysaccharide chain; i.e., an inflexibility arising from
substantially restricted rotation about the angles φ and ψ. That band
is absent in cases where increased flexibility causes cancellation of
oppositely signed CD contributions through conformational averaging.

Our approach to the assignment of the low energy saccharide CD
band is based on the theoretical work of Texter and Stevens.[75-77] We
assign the low energy band to a σ*/3s←n transition originating from the
non-bonding orbital of every oxygen atom, including the ring oxygen,
linkage oxygen and hydroxyl oxygens. The observed dichroism in the low
energy region is the result of summed contributions from each of those
transitions, each contribution reflecting the relevant conformational
average. In particular molecules the net result may be very small
dichroism. When a CD band is observed, one or the other contribution
may dominate. Assignment of the higher energy CD bands has not yet
been made.

The results we obtained on glucans and galactomannans which have
been described here are typical of the results we have obtained in
applying VUCD spectroscopy to polysaccharides. In all cases we seek
to make specific conformational determinations. Thus, in agarose[57] we
were able to monitor the course of gelation with intensity variations
in a VUCD band; in hyaluronic acid[64] we associated a strong negative
198 nm CD band with restricted flexibility of acetamido groups; in
chitins[65] we identified the trans amide conformation and intermolecular
hydrogen bonding as the important determinants of the solid state CD;
in alginates[66] we confirmed the "egg box" model of gelation; and in
chondroitins[67] we associated the additivity of CD, relative to non-
interacting monomers, with the axial disposition of the C(4) hydroxyl
group not allowing intramolecular hydrogen bonds. In each of these
cases VUCD spectroscopy proved to be a useful probe of molecular
conformation.

SUMMARY

The extension of circular dichroism measurements into the vacuum ultraviolet region provides an important additional spectroscopic tool for studying molecular conformation of biologically important compounds. Detailed and specific structural features can be extracted for even complex biopolymers. The technological development which made this possible was the development of a vacuum ultraviolet transmitting photoelastic modulator. The new technological development of accessible synchrotron radiation promises a significant improvement in the sensitivity of such measurements. A technological development for the future would be to produce circularly polarized synchrotron radiation without using an absorbing medium so that wavelengths below 130 nm can be made accessible.

ACKNOWLEDGMENTS

This work was supported in part by NIH Grant GM 24862.

REFERENCES

1. Schnepp, O., Allen, S. and Pearson, E., Rev. Sci. Instr., 41:1136 (1970).

2. Johnson, W.C., Jr., Rev. Sci. Instr., 42:1283 (1971).

3. Young, M.A. and Pysh(Stevens), E.S., Macromolecules, 6:790 (1973); Young, M.A., Ph.D. Thesis, Brown University (1974); Pysh(Stevens), E.S., Ann. Rev. Biophys. Bioeng., 5:63 (1976).

4. Brahms, S., Brahms, J., Spoch, G. and Brock, A., Proc. Natl. Acad. Sci., 74:3208 (1977); Brahms, S. and Brahms, J., J. Mol. Biol., 138:149 (1980).

5. Duben, A. and Bush, C.A., Anal. Chem., 52:635 (1980).

6. Pysh(Stevens), E.S., in "Research Applications of Synchrotron Radiation," Eds. R.E. Watson and M.L. Perlman, Brookhaven National Laboratory Study-Symposium, Upton, New York (1973) p. 54.

7. Sutherland, J.C. and Boles, T.T., Rev. Sci. Instrum., 49:853 (1980).

8. Snyder, P., Schatz, P.N., Rowe, E.M., "Natural and Magnetic Vacuum Ultraviolet Circular Dichroism Measurements at the Synchrotron Radiation Center University of Wisconsin-Madison, "refer to this vol., 43.

9. Nelson, R.G. and Johnson, W.C., J. Am. Chem. Soc., 94:3343 (1972); 98:4290, 4296 (1976).

10. Dickinson, H.R. and Johnson, C.W., J. Am. Chem. Soc., 96:5050 (1974).

11. Lewis, D.G. and Johnson, W.C., Biopolymers, 17:1439 (1978).

12. Zehfus, M.H. and Johnson, W.C., Biopolymers, 20:1589 (1981).

13. Hennessey, J.P., Johnson, W.C., Bakler, C., and Wood, H.G., Biochemistry, 21:642 (1982).

14. Sprecher, C.A. and Johnson, W.C., Biopolymers, 21:321 (1982).

15. Causley, G.C. and Johnson, C.W., Biopolymers, 21:1763 (1982).

16. Bertucci, C., Chiellini, E., Solvadori, P.A., and Johnson, W.C., Macromolecules, 16:507 (1983).

17. Causley, G.C., Staskus, P.W., and Johnson, W.C., Biopolymers, 22: 945 (1983).

18. Dougherty, A.M., Causley, G.C., and Johnson, W.C., Proc. Natl. Acad. Sci., 80:2193 (1983).

19. Bowman, R.L., Kellerman, M., and Johnson, W.C., Biopolymers, 22: 1045 (1983).

20. Brahmachari, S.K., Ananthanarayanan, V.S., Brahms, S., Brahms, J., Rapaka, R.S., and Bhatnagar, R.S., Biochem. Biophys. Res. Commun., 86:605 (1979).

21. Brahms, S. and Brahms, J.G., J. Chim. Phys., 76:841 (1979).

22. Brahms, S. and Brahms, J.G., J. Mol. Biol., 138:149 (1980).

23. Brahms, S. and Brahms, J.G., Biomol. Struct. Conform., Funct., Eval., Prac. Int. Symp., Eds. S. Ramachandran, G.E., Subramanian, and N. Yathindra, Pergamon, Oxford, 2:31 (1981).

24. Salesse, R., Combarnous, Y., Brahms, S., and Garnier, J., J. Arch. Biochem. Biophys., 209:480 (1981).

25. Brahms, S., Vergne, J., Brahms, J.G., DiCapua, E., Bucher, P., and Koller T., J. Mol. Biol., 162:473 (1982).

26. Brahms, S., Vergnes, J., Brahms, J.G., DiCapua, E., Bucher, P., and Koller, T., Cold Spring Harbor Symp. Quant. Biol., 1982, 47:119 (1983).

27. Bush, C.A., Duben, A., and Ralapati, S., Biochem., 19:501 (1980).

28. Bush, C.A., Feeney, R.E., Osuga, D.T., Ralapati, S., and Yeh, Y., Int. J. Peptide Protein Res., 17:125 (1981).

29. Bush, C.A. and Ralapati, S., ACS Symposium Series, No. 150, Ed. D.A. Brant, American Chemical Society, Washington, D.C. (1981) p. 293.

30. Bush, C.A. Ralapati, S., and Duben, A., Anal. Chem., 53:1140 (1981).

31. Ahmed, A.I., Osuga, D.T., Yeh, Y., Bush, C.A., Matson, G.M., Yamasaki, R.B., and Feeney, R.E., Cyro-Lett., 2:263 (1981).

32. Bush, C.A., Dua, V.K., Ralapati, S., Warren, C.D., Spik, G., Strecker, G., and Mantreuil, J., J. Biol. Chem., 257:8199 (1982).

33. Cowman, M., Bush, C.A., and Balazs, E.A., _Biopolymers_, 22:1319
    (1983).

34. Sutherland, J.C., Griffin, K.P., Keck, P.C., and Takacs, P.Z.,
    Proc. Natl. Acad. Sci., USA, 4801 (1981).

35. Sutherland, J.C., Keck, P.C., Griffin, K.P., and Takacs, P.Z.,
    Nucl. Instrum. Methods Phys. Rev., 195:375 (1982).

36. Sutherland, J.C. and Griffin, K.P., _Biopolymers_, 22:1445 (1983).

37. Tobin, M.B., Kuebler, N.A., and Pao, Y.-H., _Rev. Sci. Instrum._,
    37:922 (1966).

38. Samson, J.A.R., "Techniques of Vacuum Ultraviolet Spectroscopy,"
    John Wiley and Sons, New York (1967).

39. Cassim, J.Y. and Yang, J.T., _Biochemistry_, 8:1947 (1969).

40. Young, M.A. and Pysh(Stevens), E.S., _J. Am. Chem. Soc._, 97:5100
    (1975).

41. Pysh(Stevens), E.S., _Proc. Natl. Acad. Sci. (USA)_, 56:825 (1966).

42. Pysh(Stevens), E.S., _J. Chem. Phys._, 52:4723 (1970).

43. Rosenheck, K. and Sommer, B., _J. Chem. Phys._, 46:532 (1967).

44. Balcerski, J.S., Pysh(Stevens), E.S., Bonora, G.M. and Toniolo, C.,
    J. Am. Chem. Soc., 98:3470 (1976).

45. Toniolo, C., Bonora, G.M., Palumbo, M., and Pysh(Stevens), E.S.,
    in "Peptides 1976, Proc. 14th European Peptide Symposium,"
    Ed. A. Loffat,Univ. De Bruxelles, Namur, Belgium (1976) p. 597.

46. Toniolo, C. Bonora, G.M., Grisma, M., Bertanzon, F., and Stevens,
    E.S., _Makromol. Chemie_, 182:3149 (1981).

47. Kelly, M.M., Pysh(Stevens), E.X., Bonora, G.M., and Toniolo, C.,
    J. Am. Chem. Soc., 99:3264 (1977).

48. Paskowski, D.J., Stevens, E.S., Bonora, G.M., and Toniolo, _Biochim._
    Biophys. Acta, 535:188 (1978).

49. Liang, J.N., Stevens, E.S., Bonora, G.M., and Toniolo, C., in
    "Proc. Sixth American Peptide Symposium (Peptides: Structure and
    Biological Function)," Eds. E. Gross and J. Meienhofer, Pierce
    Chemical Co. (1979) p. 245.

50. Coffey, R.T., Stevens, E.S., Toniolo, C., and Bonora, G.M.,
    Makromol. Chemie, 182:941 (1981).

51. Palumbo, M., Bonora, G.M., Toniolo, C., Peggion, E., and Stevens,
    E.S., in "Proc. Fifth American Peptide Symposium," Ed. M. Goodman
    (1978) p. 399.

52. Toniolo, C., Bonora, G.M., Palumbo, M., Peggion, E. and Stevens,
    E.S., _Biopolymers_, 17:1713 (1978).

53. Chou, K.C., Pottle, M., Nemethy, G., Ueda, Y., and Scheraga, H.A., J. Mol. Biol., 162:89 (1982).

54. Chou, K.C. and Scheraga, H.A., Proc. Natl. Acad. Sci. USA, 79:7047 (1982).

55. Chou, K.C., Nemethy, G. and Scheraga, H.A., J. Mol. Biol., 168:389 (1983).

56. Balcerski, J.S., Pysh(Stevens), E.S., Chi Chen, G., and Yang, J.T., J. Am. Chem. Soc., 97:6274 (1975).

57. Liang, J.N., Stevens, E.S., Morris, E.R., and Rees, D.A., Biopolymers, 18:327 (1979).

58. Stipanovic, A.J., Stevens, E.S., and Gekko, K., Macromolecules, 13:1471 (1980).

59. Buffington, L.A., Stevens, E.S., Morris, E.R., and Rees, D.A., Int. J. Biolog. Macromolecules, 2:199 (1980).

60. Stipanovic, A.J., and Stevens, E.S., Int. J. Biolog. Macromolecules, 2:209 (1980).

61. Stipanovic, A.J. and Stevens, E.S., in "ACS Symposium Series No. 150," Ed. D.A. Brand, American Chemical Society, Washington, D.C. (1981) p. 303.

62. Stipanovic, A.J. and Stevens, E.S., Biopolymers, 20:1183 (1981).

63. Stipanovic, A.J. and Stevens, E.S., J. App. Polymer Sci., 37:277 (1983).

64. Buffington, L.A., Pysh(Stevens), E.S., Chakrabarti, B., and Balazs, E.A., J. Am. Chem. Soc., 99:1730 (1977).

65. Buffington, L.A. and Stevens, E.S., J. Am. Chem. Soc., 101:5159 (1979).

66. Liang, J.N., Stevens, E.S., Frangou, S.A., Morris, E.R. and Rees, D.A., Int. J. Biolog. Macromolecules, 2:204 (1980).

67. Stipanovic, A.J. and Stevens, E.S., Biopolymers, 20:1565 (1981).

68. Liang, J.N. and Stevens, E.S., Int. J. Biolog. Macromol., 4:316 (1982).

69. Dea, I.C.M. and Morrison, A., Adv. Carbohydr. Chem. Biochem., 31:241 (1975).

70. Frei, E. and Preston, R.D., Proc. Roy. Soc. (B), 169:127 (1968).

71. Palmer, K.J. and Ballantyne, M., J. Am. Chem. Soc., 72:736 (1950).

72. Sundararajan, P.R. and Rao, V.S.R., Biopolymers, 9:1239 (1970).

73. Sharman, W.R., Richards, E.L., and Malcolm, G.N., Biopolymers, 17:2817 (1978).

74. Morris, E.R. and Sanderson, G.R., in "New Techniques in Biophysics and Cell Biology," Eds. R.H. Pain and B.J. Smith, Wiley, London (1973) pp. 113-147.

75. Texter, J. and Stevens, E.S., J. Chem. Phys., 69:1680 (1978).

76. Texter, J. and Stevens, E.S., J. Chem. Phys., 70:1140 (1979).

77. Texter, J. and Stevens, E.S., J. Org. Chem., 44:3222 (1979).